P9-ARO-916

Liberty for All?

A HISTORY OF US

BOOK ONE	The First Americans
BOOK TWO	Making Thirteen Colonies
BOOK THREE	From Colonies to Country
BOOK FOUR	The New Nation
BOOK FIVE	Liberty for All?
BOOK SIX	War, Terrible War
BOOK SEVEN	Reconstruction and Reform
BOOK EIGHT	An Age of Extremes
BOOK NINE	War, Peace, and All That Jazz
BOOK TEN	All the People

Oxford University Press

In these books you will find explorers, farmers, cowboys, heroes, villains, inventors, presidents, poets, pirates, artists, slaves, teachers, revolutionaries, priests, musicians—the girls and boys, men and women, who all became Americans....

OXFORD
A HISTORY OF
US
BOOK FIVE

Liberty for All?

Joy Hakim

Oxford University Press
New York

Oxford University Press
Oxford New York
Athens Auckland Bangkok Bombay
Calcutta Cape Town Dar es Salaam Delhi
Florence Hong Kong Istanbul Karachi
Kuala Lumpur Madras Madrid Melbourne
Mexico City Nairobi Paris Singapore
Taipei Tokyo Toronto

and associated companies in
Berlin Ibadan

Designer: Mervyn E. Clay

Produced by American Historical Publications

Published by Oxford University Press, Inc.
200 Madison Avenue, New York, New York 10016

Library of Congress Cataloging-in-Publication Data
Hakim, Joy.
Liberty for all? / Joy Hakim.
p. cm.—(A history of US: bk. 5)
Includes bibliographical references (p.) and index.
Summary: Discusses the period of growth in American history prior to the Civil War, describing the lives of people from a variety of backgrounds, including Jedediah Smith, Emily Dickinson, John James Audubon, and Sojourner Truth.
ISBN 0-19-507753-9 (lib. ed.)—ISBN 0-19-507765-2 (series, lib. ed.)
ISBN 0-19-507754-7 (paperback ed.)—ISBN 0-19-507766-0 (series, paperback ed.)
ISBN 0-19-509510-3 (trade hardcover ed.)—ISBN 0-19-509484-0 (series, trade hardcover ed.)
1. United States—History—1815–1861—Juvenile literature.
2. United States—History—1815–1861—Biography—Juvenile literature. [1. United States—History—1783–1865.] I. Title. II. Series:
Hakim, Joy. History of US; 5.
E178.3.H22 1994 vol. 5
[E339]
973 s—dc20
[973.6] 93-19600
CIP
AC

5 7 9 8 6
Printed in the United States of America
on acid-free paper

THIS BOOK IS FOR MY FAVORITE MATHEMATICIANS,
JEFFREY HAKIM AND HANNA SANDLER

Winslow Homer, *The Cotton Pickers*

This is my letter to the World
That never wrote to me—
That simple News that Nature told—
With tender Majesty

Her message is committed
To Hands I cannot see—
For love of Her—Sweet countrymen—
Judge tenderly—of Me

—EMILY DICKINSON

The pony rider was usually a little bit of a man, brimful of spirit and endurance. ...He rode a splendid horse that was born for a racer and fed and lodged like a gentleman; kept him at his utmost speed for ten miles, and then, as he came crashing up to the station where stood two men holding fast a fresh, impatient steed, the transfer of rider and mailbag was made in the twinkling of an eye, and away flew the eager pair and were out of sight before the spectator could hardly get the ghost of a look.

—MARK TWAIN, *ROUGHING IT*

Oh! isn't it a pity, such a pretty girl as I—
Should be sent to the factory to
pine away and die?
Oh! I cannot be a slave,
I will not be a slave,
For I'm so fond of liberty
That I cannot be a slave.

—SONG OF THE LOWELL MILL GIRLS ON STRIKE IN 1836

No man can put a chain about the ankle of his fellow man without at last finding the other end fastened about his own neck.

—FREDERICK DOUGLASS

Contents

Henry David Thoreau

PREFACE
Antebellum: Say *Aunty Belle* and Add *um*

In Latin, **ante** *means "before" and* **bellum** *means "war," and, since many 19th-century Americans knew Latin, they used that word,* **antebellum** *(say aunty-BELL-um), to describe the time before the Civil War—the terrible conflict that divided the nation and then brought it back together again. As you'll soon see, slavery was the most important issue in those days—and it was an ugly issue—but it wasn't the only thing on people's minds. There was gold, real gold, to be dug; there was land to be explored; there was war with Mexico and wars with the Native Americans; there were roads to be built; and industries to be started; and more, lots more.*

Andrew Jackson was a man of the people who brought a new kind of democracy to the presidency.

The subject of a debate at Harvard College in 1828 was: *Can one man be president of the United States when it is eventually settled from Atlantic to Pacific?* The answer of the winning team was *no*, which, of course, was the only sensible answer.

How could one person be president when there was no way for the president in Washington, D.C., to communicate with the West Coast, except perhaps by ship around Cape Horn? Only a few explorers had been across the country, and no one had even tried to make the trip with a wagon or coach. If they had tried they would have found it close to impossible.

In 1828 there were neither telegraphs nor telephones nor railroads nor highways. They were on their way. The 19th century would see changes—enormous and rapid changes—but in 1828 only the very farsighted had an inkling of what technology would bring.

That year of 1828 was an election year, and the turnout of voters—to

When Andrew Jackson declared war on the Bank of the United States (you'll read about that on page 11), people who disagreed with him said the president just wanted glory for himself. This cartoon shows him on a mock bank bill (called a *shinplaster*).

make Andrew Jackson president—was astounding. In 1824, only 356,038 Americans had bothered to vote in the presidential election. Four years later, partly because of new voting laws, 1,155,340 white men went to the polls—an increase of 224 percent.

Jackson took that word *democracy*, which scared some people, and glorified it. It became the essential word in American politics. Jackson called his presidency a revolution—as Jefferson had called his presidency—and they were both right. Each believed in democracy, but Jackson made it happen. He made people's government—democracy—respectable.

Elections (above, a county election) were rowdy affairs, but the 1828 presidential campaign was especially nasty. Jackson's men painted John Quincy Adams as an aristocrat who spent taxpayers' money on "gambling equipment" for the White House (actually a chess set and a billiard table).

Who were these Americans who had suddenly decided to go to the polls and take part in running their own government?

Back in 1800 there were 15 farmers for every person who lived in a city or town. By Andrew Jackson's time—28 years later—there were 10 farmers for every town dweller. The yeoman farmer—Jefferson's ideal American—was on his way to near-extinction. But that wasn't clear to Andrew Jackson, or anyone else. What was clear was that this was a nation of optimistic and restless peoples. By 1840, one third of our citizens had moved west of the Appalachian Mountains. Ohio had become the third largest state in population, behind only Pennsylvania and New York. If there was opportunity, we Americans would move. And the West seemed to be beckoning.

Those self-sufficient yeoman farmers, who had hardly seen or needed money, were now growing wheat and cotton: cash crops. The farmers were using the money they got for their cotton and wheat to buy food for their families—and luxuries too. America was developing a money economy, and that was causing all kinds of discussion. Should a farmer borrow money and go into debt to buy farm equipment? Was that immoral, or was it just smart?

Immoral **means wicked. Borrowing money, traditionally, was a sign of trouble. In the old farming world, where money was scarce, to go into debt (especially if you didn't really have to) seemed immoral and irresponsible. But, in the new commercial world, borrowing was often necessary and reasonable if a business, or farm, was to grow.**

Money was changing the United States (and England and other nations, too). The Industrial Revolution started it. It made capital—money—essential for business growth. New commercial interests (businesses and railroads and banks) were looking for money and demanding power and influence. The old guard (those who already had money and influence) was trying to control and slow the rate of growth.

It was all confusing and terrifying to those who were attached to the new idea of self-government. John Calhoun wrote that "liberty was never in greater danger." Andrew Jackson was the common person's president. He thought the money interests were asking for special privileges from the government, and getting them. "If we cannot at once...make our government what it ought to be, we can at least take a stand against....the advancement of the few at the expense of the many."

Jackson took his stand when he killed the Bank of the United States. That helped set off a depression—but it was Jackson's successor, Martin Van Buren, who had to deal with it. Yet nothing was going to stop the drive for wealth and achievement. Not in a free nation.

Nor could anything stop people from getting angry about it. When the writer James Fenimore Cooper came home in 1833, after a long stay

Unlike the cartoon of Jackson on page 9, this was a real shinplaster—an actual 50-cent bill put out by a local savings bank.

Andy Jackson: Big Bank Buster

The Bank of the United States was chartered by Congress, and the federal treasury was its biggest depositor. It helped regulate currency and the economy in general. The bank's charter had to be renewed regularly. Jackson picked on the Bank of the United States as a symbol of aristocratic privilege and influence; he was convinced that its directors stood against the idea of democratic self-government. (Actually, the bank had helped make the country economically sound, but—no question about it—it did favor older commercial interests over farmers and the new business people.) Jackson called the bank a "monster" and a "hydra of corruption"; he made the bank *the* big issue when he ran for reelection. Then he set out to destroy the bank, and did. But he didn't replace it with anything, so small banks were free to do what they wished—and that led to monetary confusion and disorder. It was a mixed victory. The common people gained power, but America's money system was never again as sound as when the bank was in control.

Most Americans didn't understand the complex world of economics and banking. The bank battle was actually a fight between the old, aristocratic leaders (who had money) and the new manufacturers, builders, and promoters (who needed money to invest and wanted as little restraint as possible). The old guard supported the bank, hated Jackson, and founded the anti-Jackson Whig Party. Politically, it was a hot and angry time.

To most voters, Jackson was the champion of ordinary people against the money kings.

Jefferson may have had Aristotle's idea about happiness in mind when he wrote the Declaration of Independence. Aristotle, a Greek philosopher, thought happiness could be found in a thoughtful life.

in Europe, he was horrified by an attitude that seemed to have taken hold in his country. "The desire to grow suddenly rich has seized on all classes," one character complained in Cooper's novel *Homeward Bound*. Were Americans losing their souls in a search for riches? Some thought so.

But that drive for individual riches had surprising consequences. It would make the whole nation rich. It provided better food, clothing, and shelter to more people than in any nation before. It spawned ideas, and poetry, and songs.

You see, this new nation had an unusual goal: it had been directed by its founders to pursue happiness. That was such a simple, logical goal—but no nation seems to have thought of it before. Besides, what is happiness, and how do you pursue it?

There were lots of answers to that question. Some people said happiness was having land. Some said it was being rich. Some said it was going to school. Some said it was having a fine family. Some said it could be found in religion. Some said it was being free. Some said...well, you get the idea: different people have different ideas about what will make them happy.

For most Americans, happiness meant the hope of owning land, and in the 19th century wave upon wave of settlers pushed westward, seeking land and success. (Of course, their quest often meant disaster for the first Americans—the Indians.)

This was a nation trying to find ways to let each citizen pursue his own goals—without hurting anyone else—which was a new thing for a nation to do, and not easy at all. Some people were clearly off the track. How can anyone pursue happiness if he or she is a slave, or is keeping others enslaved? How can anyone be happy if he is being lied to, or is thrown off his land?

This country was very young. We had a lot to learn. The antebellum time was a period of growth and experimentation. It was a time when gold seekers took up pickaxes and shovels, when preachers led big outdoor church gatherings, when abolitionists tried to right a terrible wrong, when some people were vile and others were virtuous. Whether it all seemed mostly good or mostly bad, or something in between, one thing was sure: it was not a time to be bored.

1 The Long Way West

The states carved from Spanish holdings are: Texas, California, Nevada, Utah, New Mexico, Arizona, and parts of Wyoming and Colorado.

The states from the Louisiana Purchase territory are: Arkansas, Iowa, Louisiana, Missouri, Montana, Nebraska, North Dakota, South Dakota, Oklahoma, and parts of Colorado, Kansas, Minnesota, and Wyoming.

The states from the Oregon Territory are Idaho, Oregon, and Washington.

After his Rocky Mountain expedition, the explorer Stephen Long's main claim to fame was surveying the route for the first U.S. railroad, the Baltimore & Ohio.

When President Thomas Jefferson purchased the Louisiana Territory from France, in 1803, a few people grumbled, but most Americans approved. Hardly anyone had liked the idea of France having land on the border of the United States, or of France controlling the port of New Orleans. Otherwise, nobody seemed to know what to do with the territory. It was a huge hunk of land that went from the Mississippi River to the Rocky Mountains. (Later it was turned into 13 states.) But the Louisiana Purchase didn't include southern and western lands that stretched from Texas to California and north to the edge of Wyoming (a region that would become all or part of eight more states). That land was all called California, and Spain still owned it. And it didn't include a northern region known as the Oregon Country. Great Britain was claiming Oregon—although there was some dispute about it.

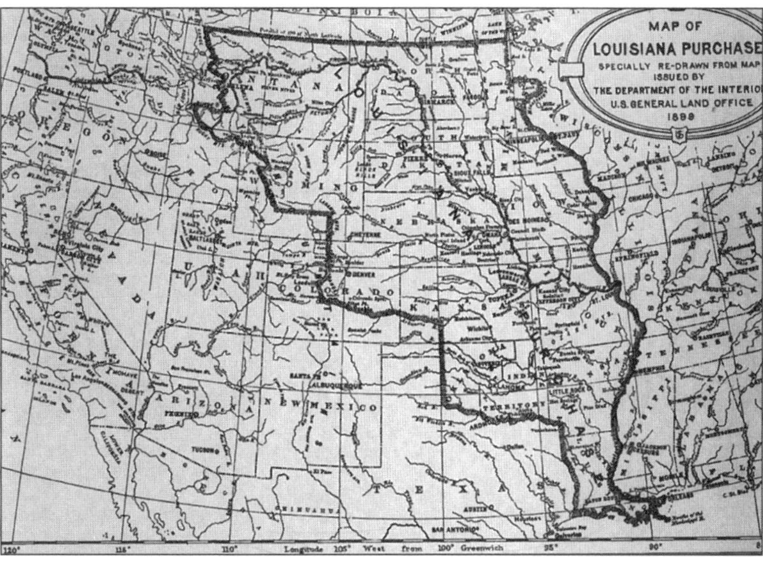

Actually, it was mighty strange that France, Spain, England, and the United States were buying and selling the land, as it was inhabited, and presumably owned, by Indian

Until Long reached Council Bluffs in the ship *Western Engineer* (sketched above by one of the expedition's official artists, Titian Peale), no steamboat had gotten so far up the Missouri River. The expedition wintered at Council Bluffs and then went on by canoe up the Platte River to the Rockies.

It was 1839 when the first photographs were taken in America (by Samuel F. B. Morse, whom you'll hear more about in this book, and John W. Draper). The subject had to sit still for half an hour because that was how long it took to get a proper exposure.

nations that had lived there for thousands of years. But most Europeans and Americans talked of the Indians as "savages," and they usually acted as if those savages had no rights. Of course, the Indians believed they had rights. They weren't willing to give up their land. They would fight for it; but that came later, when railroad trains—called "iron horses"—thundered across the buffalo ranges, and an endless parade of men from the East killed the buffalo and the Indian's way of life, too.

Early in the 19th century, most people in the United States thought there was plenty of room for the Indians on the land to the west of the Mississippi. And, at first, the Indians were friendly toward the newcomers. Besides, few Americans seemed to have any desire to live in the Louisiana Territory. Mostly, they were just curious about it—the way we wonder about distant planets. Besides, if you believe you own something, it makes sense to find out exactly what it looks like, and—aside from what Lewis and Clark and a few other explorers had reported—no one knew much about the western lands.

So, in 1819, President Monroe's secretary of war, John C. Calhoun, decided to organize an exploring expedition to find out about that across-the-Mississippi region. He asked a U.S. army major, Stephen H. Long, to take charge. (Long had been an assistant professor of mathematics at West Point after graduating near the top of his class at Dartmouth College.) He set out on a two-year journey. Edwin James, who was a scientist and a surgeon, went with him; so did artists Titian Peale and Samuel Seymour. They were sent to draw pictures of the landscape, the animals, and the flowers. (Those were the days before photographers.) James wrote a two-volume account of the journey and made an atlas, too.

None of the members of the expedition seemed particularly impressed with what they found. This was what Long said about the plains that went from the edge of the Mississippi woodlands to the Rocky Mountains:

> *In regard to this extensive section of the country, I do not hesitate in giving the opinion, that it is almost wholly unfit for cultivation, and of course, uninhabitable by a people depending upon agriculture for their subsistence.*

He called it the "haunt of bison and jackal," whose "sole monarch" was the prickly pear. On Edwin James's official map of the expedition, the Great Plains is labeled "The Great American Desert."

It was just as well, thought Major Long, that this land was unfit for human habitation (except by Indians, and most people didn't worry about them). It would, he said, *serve as a barrier to prevent too great an extension of our population westward.*

This was actually fertile land, but Easterners didn't know it. They were used to chopping down trees to make farmland. People listened to Long, and, for the next 20 years, most travelers who ventured west agreed: the Great Plains was an uninhabitable desert. So, aside from the Native Americans who lived there, only a few traders, trappers, mountain men, and scientists headed for the Great Plains—or beyond, to the tall, snow-capped mountains.

The Sioux, the Pawnee, the Osage, the Arapaho, the Kiowa, the Comanche, and the Wichita were some of the Indian peoples living on the land explored by the Long expedition.

A ***prickly pear*** is a cactus with an egg-shaped, edible fruit.

Below, Long's company camped among a group of Kiowa Indians. The expedition's flag shows a white and a brown hand, clasped, and the symbols of power and peace: a sword and a pipe.

The artists on Long's expedition brought home some of the first sketches of life west of the Mississippi since Lewis and Clark had returned 15 years earlier. Above, three Indians from different tribes, including an Arapaho on the right.

2 Mountain Men

TO

Enterprising Young Men.

THE subscriber wishes to engage ONE HUNDRED MEN, to ascend the river Missouri to its source, there to be employed for one, two or three years.—For particulars, enquire of Major Andrew Henry, near the Lead Mines, in the County of Washington, (who will ascend with, and command the party) or to the subscriber at St. Louis.

Wm. H. Ashley.

February 13 —98 tf

Ashley made a deal with the mountain men: he transported and outfitted them for a year, and in return they gave him half the year's catch.

The ad in the St. Louis newspaper, in February 1822, called for men willing to try something new. Instead of buying furs from the Indians—as most American fur traders did—they were to live as the Indians lived and trap and hunt furs themselves. They were to go out into the unexplored West—to the Rocky Mountains and beyond—with a gun and a knife, some coffee and flour, to eat the game they shot and the berries they found, and to trap beaver for the fur market.

Rendezvous—say RON-day-voo—it means a meeting or get-together agreed on in advance.

William Ashley, who put the ad in the paper, would arrange a once-a-year riverside meeting—a month-long rendezvous—where they could all get together, sell their beaver, feast, race on horseback, wrestle, trade stories, and have a fine time. If they wanted to, they could stay away from civilization for years. They would live with grizzly bears, rattlesnakes, mountain lions, blizzards, floods, and drought. They would share the land with the Native Americans, who would sometimes teach them the ways of the mountains and—sometimes—kill them.

Davy Crockett's Almanack **told tall tales about mountain men and the frontier, excitingly illustrated. It had nothing to do with Crockett himself.**

To show their fearlessness, the mountain men wore their hair long. It proved they were not afraid of being scalped.

Some people like danger and adventure, some like to be free of civilization, and some like to live by their wits. It was those special people who headed west.

In the year 1832, a thousand men turned up for the rendezvous. Their skins were white, black, and copper, and they all got along, at what some say was the best party the West has ever seen.

If only Daniel Boone had been around—he died in 1820 at age 86—Boone would have loved being one of Ashley's Mountain Men. That was what they called themselves—mountain men—and they became a kind of brotherhood. Legends were told of them: of Jim Bridger, Jedediah Smith, James Beckwourth, Tom Fitzpatrick, and others.

A German fur trader (named Frederick Adolph Wislizenus) went to a rendezvous and wrote this of the mountain men:

> *In small parties they roam through all the mountain passes. No rock is too steep for them; no stream too swift. Withal, they are in constant danger from hostile Indians, whose delight it is to ambush such small parties, and plunder them, and scalp them. Such victims fall every year....But this daily danger seems to exercise a magic attraction over most of them. Only with reluctance does a trapper abandon his*

The mountain men set beaver traps in streams. They smeared the bait with *castor*, taken from a beaver's musk glands, and tied the trap to a pole. They drove the pole into the mud by the bank, near fresh prints that showed beaver were around.

Audubon's *American Beaver*. More on Audubon in chapter 29.

For a time beaver hats were so fashionable that in 20 years beavers were almost extinct.

dangerous craft; and a sort of serious home-sickness seizes him when he retires from his mountain life to civilization. In manners and customs, the trappers have borrowed much from the Indians. Many of them, too, have taken Indian women as wives. Their dress is generally of leather. The hair of the head is usually allowed to grow long. In place of money they use beaver skins.

Jedediah Smith looked like the other mountain men; he wore the same buckskin clothes, fringed at the seams, with buffalo-hide moccasins on his feet. Indian clothes they were, and if you wanted to com-

EXPEDITIONS of Jedediah Smith

1 Smith sets out for "Fur Country"
2 Attacked by grizzly bear. Friend sews scalp and left ear back on
3 Follows Crow directions and "discovers" the South Pass.
4 First trappers' rendezvous
5 Returns to St. Louis with boatload of furs
6 Second rendezvous
7 Out of food; lives on insects and rodents
8 Mohave Indians help party cross the desert
9 Attacked by Paiute Indians
10 Trapped by deep snow
11 Crosses the Great Basin in 32 days with little food or water
12 Third rendezvous
13 Smith returns; Mohave Indians kill 10 of party
14 Sails to San Francisco
15 Party jailed under suspicion of being American spies
16 Jailed again. Escapes with 250 horses and mules
17 All but two of Smith's party massacred
18 Smith retires
19 Killed by Comanches on a business trip to Santa Fe

Samuel Woodhouse was a scientist who explored Indian territory in what is now Oklahoma. He and the mountain men endured the same hazards—Woodhouse lost the use of a hand from a rattlesnake bite.

pliment a mountain man you could say you mistook him for an Indian.

But Jedediah Smith was different. He didn't drink, swear, or chew tobacco. His Bible was as important to him as his rifle. He carried it with him always and he knew most of its stories by heart. It was Smith who found the South Pass, a gap through the Rockies in the Wyoming region. He understood its importance. That pass (like the Cumberland Gap, where Daniel Boone cut a trail through the southern Appalachians) was a place settlers could use to get wagons through the mountains.

Jed Smith was guided by a feeling that what he was doing was important, that he was helping the nation grow. In a letter to his brother, he wrote:

> *It is, that I may be able to help those who stand in need, that I face every danger—it is for this that I traverse the mountains covered with eternal snow—it is for this that I pass over the sandy plains, in heat of summer, thirsting for water, and am well pleased if I can find a shade, instead of water, where I may cool my overheated body—it is for this that I go for days without eating, and am pretty well satisfied if I can gather a few roots, a few snails, or, much better satisfied if we can afford ourselves a piece of horse flesh, or a fine roasted dog.*

The South Pass had been found in 1812 by fur trappers—but it was then forgotten. Jed Smith rediscovered it in 1824.

Jedediah was just getting started as a mountain man when he survived his first Indian massacre. Fifteen others didn't. A few years later, in 1826, Smith led a group of trappers across deserts and mountains toward California and the Pacific coast. There were two ways to get to California. You could go over the killer peaks of the Sierra Nevada mountains, or you could try the hot, parched, impossible desert. No white man had ever gone overland from the United States to California (which was then part of Mexico). When Mexican officials saw the bearded, wild-looking hunters (who had trekked across the desert), they didn't believe they'd come from the East; they didn't think anyone could make it. They decided they were spies and arrested them. Smith talked his way out of jail. He was told to leave the territory. He did. He headed back east. This time he went across the mountains; he got back in time to catch that year's rendezvous.

Titian Peale, of the Long expedition, painted these grizzlies.

Before long he was ready to try for California again. When he did,

Jim Beckwourth couldn't read, but his story got written down. It is called *The Life and Adventures of James P. Beckwourth.*

The biggest petrified forest is in Arizona; the most famous hot geyser, Old Faithful, is in Yellowstone National Park in Wyoming.

Later on, Jim Bridger guided settlers, explorers, and surveyors. He was, said a historian, "an atlas of the West."

Indians killed 10 of his men and captured all his horses. That didn't stop Jed Smith. He made it to California and went up the coast to Oregon. There, more Indians attacked; only Jed Smith and two others escaped.

Then there was the grizzly bear who tried to scalp Jedediah. Jed pulled his head out of the grizzly's mouth, but half his ear got ripped off. One of his men sewed the ear back and a few days later Smith was on the trail again. Finally, in New Mexico, in 1831, Jed Smith's luck ran out. Comanche Indians are said to have caught him. His body was never found.

The mountain men and the Indians—sometimes they were the best of friends. Often they were mortal enemies. Consider James Beckwourth: he had been a slave, and at 20 had run away and made it to the mountains. Beckwourth had an Indian wife and was adopted by the Crow and became a chief. He learned to use a tomahawk as well as any Indian. But, like most of the mountain men, Beckwourth was restless. He didn't stay an Indian for long. He went off to California, where he discovered a pass through the Sierra Nevadas. That pass still bears his name.

Everyone who knew the mountain men had stories to tell of Jim Bridger; he was always getting in and out of scrapes. Actually, they didn't have to tell stories about Bridger; he told the best stories himself. He told of a petrified forest, where trees had turned to stone, and that story was true. Then he told of petrified birds who sang petrified songs and how he crossed over a petrified canyon without a bridge, because the law of gravity was petrified there. When he told of springs he had found whose water came out of the ground so hot it would cook your food—well, no one believed that, even though it was so. It was Bridger who discovered the Great Salt Lake; for a while, because it was salty and huge, he thought he'd found the Pacific Ocean.

Once, sitting around a campfire, an army officer told Bridger a story from Shakespeare. Bridger liked the story so much that when he learned there was a set of Shakespeare's plays on a wagon train, he bartered a pair of oxen for the books, although he had never learned to read. Then he hired a boy to read them to him. After that he told Shakespeare's stories in his own words.

But that was after 1840, when the rendezvous was finished and the beaver trade too. In Paris they were wearing silk, not beaver, hats. Besides, the beaver were mostly gone, from overtrapping. The mountain men became guides, taking wagon trains of pioneers across the mountains and deserts whose trails only they and the Indians knew. Now they had to deal with people, instead of rattlesnakes and bears—some of the mountain men preferred rattlesnakes.

3 Riding the Trail to Santa Fe

Only seven years after he discovered Pikes Peak, Zebulon Pike died fighting the British in the War of 1812.

Back in 1806, a young army officer named Zebulon Montgomery Pike headed west from St. Louis on an exploring mission. It was the same year Lewis and Clark returned from their trip. They had gone northwest; Pike went southwest.

Traveling along the Arkansas River into Colorado, he came upon a towering mountain that loomed straight up from the level plain. Pike tried to climb it, failed, and wrote: *no human being could have ascended to its pinical.* (Which means: no one can climb that peak!)

Fourteen years later, three human beings from Stephen Long's expedition proved to be better climbers (and better spellers) than Zeb Pike. They ascended to the pinnacle. By that time Zebulon Pike was famous. That was because he had written about his adventures. People were fascinated with his account, especially with his description of an enchanting—and rich—little Spanish town named Santa Fe. The mountain Pike never climbed was named Pikes Peak in his honor—and traders began itching to get to Santa Fe, that rich Spanish town. After Long came back with his report, and the mountain men's stories got circulated, some traders headed for northern Mexico.

The first of them got thrown in jail. The Spanish-ruled Mexican territories didn't welcome outsiders. Santa Fe stayed isolated—and tempting.

In the meantime, Americans were moving west. In 1821, Missouri became a state. In September of that year, William Becknell led four men west from Old Franklin, Missouri, "for the purpose of trading for

Julia Archibald Holmes climbed Pikes Peak in 1858. As far as we know, she was the first woman to do so. More on Julia Holmes in chapter 22.

In the United States there is more space where nobody is than where anybody is. This is what makes America what it is," wrote a 20th-century American writer named Gertrude Stein. Is that true today? Don't answer that question too quickly. Look at a map.

horses and mules and catching wild animals of every description." Becknell and his companions had goods loaded onto mule backs. They planned to trade with Indians. They weren't having much luck, when they met a group of Mexicans who urged them to go to Santa Fe. Mexico had just become independent of Spain; perhaps the new governor would let them into the territory.

Becknell reached Santa Fe in November, quickly sold everything he had for Spanish silver dollars, packed the coins in bulging rawhide bags, and was back in Missouri 48 days later. He brought a message from Governor Facundo Melagres of New Mexico. American traders were now welcome in Mexican territory, said the governor. That was all that merchants in the States had to hear. Their wagons were soon cutting deep ruts in a trail west. It was called the Santa Fe Trail.

We are caraing on a smart Trade with St. tefee, wrote another poor speller—a Missouri merchant—in 1824. Becknell went back that same year. This time he piled $3,000 worth of trade goods into huge, heavy wagons. And he blazed a southern cutoff that avoided the steep mountain passes. It was a journey of more than 800 miles—through a

Josiah Gregg was a Santa Fe trader. In 1844 he published a book about the traders' life, *Commerce of the Prairies.* This picture of the "march of the caravan" was one of the illustrations.

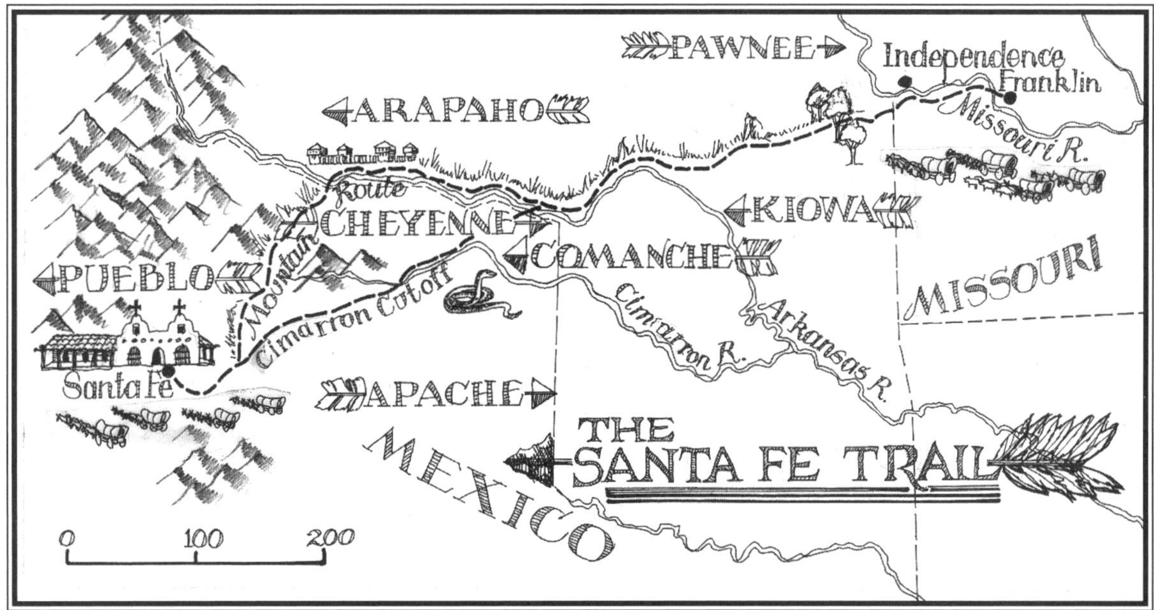

long, waterless desert region—but, with water and food in his wagons, Becknell made it. He also made a 2,000 percent profit on his merchandise. When American merchants learned that, they began pouring into Santa Fe. Some New Mexicans—especially those who had been Americans but were now Mexican, and liked the gracious, leisurely life—urged the governor to stop the trade. It was too late. There was no stopping the Americans.

Mostly it was men who traveled the Santa Fe trail. They went in big caravans. They didn't plan to settle. They were going to get rich, or for adventure, or to see new lands. Some of them did all those things. Their heavy wagons were pulled by teams of mules or oxen and were filled with cotton cloth, tin cups, socks, mirrors, cutlery, ribbons, buttons, glassware, ink, hats, gloves, and silk shawls. They brought the food that they ate on the trail, and that included cattle, chickens, and hogs. They slept in tents, or in the open air, or under the wagons.

It was a tough journey from Missouri to New Mexico, and it took over a month. The traders faced blazing heat, fierce thunderstorms,

Wagons on the Santa Fe trail had to skirt acres of "dog towns," the home of colonies of prairie dogs. Wagon wheels got stuck in the dog holes. Besides, prairie dogs taste awful, so they were no use for food.

There weren't many ambushes on the trail, but the fear of ambush was constant.

Who was Marco Polo?

A Frenchman named Alexis de Tocqueville visited the United States in the early 19th century. He was interested in studying our democracy. He wrote, "Nothing struck me more forcefully than the equality of conditions." When the open land was gone, would conditions become less equal?

maddening mosquito attacks, rattlesnakes, long thirsty stretches, and, sometimes, when they were invading Indian territory, ambush. But most of the traders were young, and they loved the adventure of it all. They thought of themselves as modern-day Marco Polos. So did others. The editor of the *Missouri Intelligencer* wrote in 1830:

> *The accounts of these inland expeditions remind one of the caravans of the East. The dangers which both encounter—the caravan of the East and that of the West—are equally numerous and equally alarming. Men of high chivalric and somewhat romantic natures are requisite for both.*

It wasn't as dangerous as the storytellers made it sound, but it was dangerous. According to a 20th-century writer, Paul Horgan, "Early encounters with Indians were peaceful; but it was not long until traders, regarding the Indians as inferior creatures, abused them; and Indians replied with all their ancient skill in savage warfare." During the first 10 years of its existence, only eight men died on the Santa Fe Trail. (It was ironic that one was Jedediah Smith. As you know, Jed had survived really dangerous trails, and he was a friend of most Indians.)

The caravans were filled with a mixture of characters. One letter writer told of a wagon train that had men of "seven distinct nations, each

Once Pikes Peak was conquered, everyone wanted to climb it. Then, 30 years later, gold was discovered nearby, and the miners arrived—"Pikes Peakers."

speaking his own native tongue." They included a talkative Frenchman who threw his hands around "with curious gesticulations," two "wanderers from Germany," two Polish exiles of "calm eccentricity," a Creek and a Chickasaw Indian, and Americans who "were mostly backwoodsmen, who could handle the rifle..."

Josiah Gregg went west with 100 wagons in 1831. When he reached New Mexico he found a world that seemed strange and exotic. Gregg had not seen adobe buildings before. He called them "heaps of unburnt bricks—nevertheless they are houses." Here is how Gregg described the scene when a wagon train pulled into Santa Fe:

> *The arrival produced a great deal of bustle and excitement among the natives.* "Los Americanos!"—"Los carros!"—"La entrada de la caravana!" *were to be heard in every direction; and crowds of women and boys flocked around to see the newcomers; while crowds of* léperos *hung about, as usual, to see what they could pilfer. The wagoners were by no means free from excitement on this occasion....they had spent the previous morning in rubbing up; and now they were prepared, with clean faces, sleek-combed hair, and their choicest Sunday suit....[Each wagoner had tied] a bran new cracker to the lash of his whip; for on driving through the streets and the* plaza pública *every one strives to outvie his comrades in the dexterity with which he flourishes this favorite badge of authority.*

When the artist who painted this picture of Santa Fe was making a sketch for it in 1866, an "exceedingly rough-looking fellow" tried to make him sell the painting at gunpoint. The artist told him it would cost $10,000—at which the man fell silent and walked off.

You know who los Americanos *are; can you figure out what* los carros, la entrada de caravana, léperos, *and* plaza pública *mean?*

Pilfer means to steal in a small way. We now spell ***bran-new*** with a *d*—*brand-new*—but it came from the bran used to ship new goods in barrels.

4 Susan Magoffin's Diary

When the Magoffins hit the trail, they took 14 wagons, 2 horses, 11 mules, and 200 oxen. Susan made the trip from Independence to Santa Fe several times; she died of yellow fever at age 27.

Only a few American women had traveled the Santa Fe Trail in 1846 when Susan Magoffin headed west from Missouri. She was 18, newly married, pregnant, and very much in love. She was excited by the adventure. *Oh this is a life I would not exchange for a good deal!* she wrote from her tent on the trail.

> *There is such independence, so much free uncontaminated air, which impregnates the mind, the feelings, nay every thought, with purity. I breathe free without that oppression and uneasiness felt in the gossiping circles of a settled home.*

Susan's husband was a wealthy trader in charge of a big wagon train. Susan traveled in a carriage with a servant. Still, it wasn't an easy trip for a proper American girl who wore long dresses buttoned to her neck. It was hot on those open plains. Magoffin was a good sport. She pitched in and fed the chickens, cooked meals, sewed, wrote in her diary, made notes on the wild flowers and animals, gathered berries, and settled arguments.

Whatever happened—and a lot did happen—she tried to keep her cheerful spirit. When the rains came and her bed was an island in the tent, she pretended it was a boat. When thunder and lightning raged and the tent collapsed, she made the best of it. When the carriage crashed down an embankment, Susan was hit on the head, knocked out, and almost killed. Somehow, she made the best of that too.

Susan Magoffin had read Josiah Gregg's diary. She thought people

The buffalo were the first to blaze the Santa Fe Trail. Josiah Gregg said where the buffalo had been it seemed like an "immense highway."

might read hers. She filled it with details, especially of people. The Indians and the Mexicans were fascinated with her and the clothes she wore.

> *A parcel of Indians are around the tent peeping in at me and expressing their opinions. It is a novel sight for them. These are the Pueblos or descendants of the original inhabitants—the principal cultivators of the soil—supplying the Mexican inhabitants with fruits, vegetables, etc.*

When she finally reached Santa Fe, she wrote:

> *What a polite people these Mexicans are…this morning…a little market girl…came in and we had a long conversation on matters and things in general, and I found that though not more than six years old she is quite conversant in all things. On receiving her pay she bowed most politely, shook hands with a kind* "adios" *and* "me alegro de verte bien" *(I am glad to see you in good health), and also a promise to return tomorrow.*

While Susan Magoffin was traveling the Santa Fe Trail, her brother-in-law James Magoffin was on a secret mission. A mission for President

"We have much pulling through sand," wrote Susan. "We stopped earlier tonight, on the bank of the Rio del Norte—it resembles the Mississippi much, muddy and dark, the banks are low, with no trees—we are buying wood every day."

Marching to Santa Fe in 1846, a soldier described the desert: "Dreary, sultry, desolate, boundless solitude reigned as far as the eye could reach....We suffered much with the heat, and thirst, and driven sand—which filled our eyes, and nostrils, and mouths, almost to suffocation."

James K. Polk. James Magoffin was a smart, successful trader who had married a Mexican woman and settled in Santa Fe. He was full of stories, and fun to be around. No one knows for sure what he was told to do—it was a *secret* mission—but he went into Santa Fe ahead of an American army and persuaded the New Mexican governor (who happened to be his brother-in-law) not to fight. Then Colonel Stephen Watts Kearny, of the United States Army, arrived and captured New Mexico without spilling any blood. New Mexico became United States territory. Some of the Spanish-speaking New Mexicans were furious. They felt they were being taken over by foreign invaders. And they were. (Those New Mexicans didn't seem to remember what *they* had done to the Indians.) But many New Mexicans were happy to be under U.S. rule.

"New Mexicans," said Col. Kearny when he reached Santa Fe, "we do not mean to murder you or rob you of your property....My soldiers will take nothing from you but what they pay you for. In taking possession of New Mexico we do not mean to take your religion away from you."

United States control didn't change life a whole lot in Santa Fe. It did make things easier for the Santa Fe traders. They didn't have to fill out the long, tiresome papers the Mexican officials had demanded.

In 1866 (which was after the Civil War), trail traffic hit its peak. That year, 5,000 freight wagons headed west from Missouri. Stagecoaches were making regular runs on the trail, too. By then, railroad trains had already chugged into eastern Kansas. When their tracks reached New Mexico, in 1879, the Santa Fe Trail was history.

The New Americans

It is 1846 and a disease—a blight—has destroyed Ireland's potato crop. There is almost nothing to eat for most of the Irish. It is hard for us to imagine people starving to death, but that is happening in Ireland. It is called the Great Potato Famine. In addition, a "poor law" is taxing small farmers much more than it taxes the rich. Between 1847 and 1854, 1.6 million Irish come to the United States.

Germany is having problems, too. New factories are putting people out of work; cities and villages are filled with poor people; German farmers aren't doing well; and a freedom revolution fails. Then a rumor spreads through Germany: America is going to close its doors. Immigrants will no longer be allowed. It is just a rumor—it isn't true—but it

Are Called Immigrants

starts a panic. Many Germans pack their bags and hurry off to the land of freedom and opportunity.

In China there are few jobs. In America railroads need building. So, beginning in the late 1840s, boatloads of Chinese come, work on the railroads, and stay.

All these people are ambitious, or they wouldn't have made the big journey to the "New World." They are the kind of people who don't mind moving—and moving again once they arrive in America.

Between 1845 and 1860, more immigrants come to this country in proportion to the total population than at any other time in our history. (Be sure you understand: there will be a larger number of immigrants later, but there will also be a much larger total population.) They come from Germany, China, Ireland, India, Norway, Sweden, Finland, Poland, Russia, Italy, Greece, Ethiopia, Morocco, Japan, and Turkey.

Have you ever heard of the Isle of Man? It is near England. People from the Isle of Man are called "Manx." Manx come to the United States. Most of the immigrants come from northern Europe, but wherever there is a nation, there is usually someone who sets out for America.

Rich, successful people don't leave their homelands. Why should they? The people who come to America are mostly poor, or troubled, or persecuted, or kidnapped, or adventuresome. Some of them stay in the East, where the ships leave them, but many travel on—to the West. In America, those rag-tag, adventuresome people will show the world the power of opportunity.

Not all these Irish emigrants find a better life in America. Many die young, worn out by domestic or construction work. But all of them can hope.

5 Pioneers: Taking the Trail West

I've often been asked if we did not suffer with fear in those days but I've said no we did not have sense enough to realize our danger we just had the time of our lives but since I've grown older and could realize the danger and the feelings of the mothers, I often wonder how they really lived through it all.

—NANCY HEMBREE SNOW BOGART, WRITING OF LIFE ON THE OREGON TRAIL

Q: I am food and will walk with the pioneers and fill their cups along the way. What am I?

A: a cow.

Camping under a cottonwood tree at the Missouri ford, Council Bluffs. Cottonwoods were important to settlers in the almost treeless prairie, because they grew very fast.

Their ancestors had hugged parents and grandparents, wiped away their tears, and set out for a New World. Now another generation of men, women, and children was heading out toward a little-known world. They, too, were leaving parents and grandparents—often never to see them again. They were heading west. For some it would be a great adventure; some would not live to finish the journey. They were going for the reasons that usually make people move: because they wanted a better life for themselves and their children, or because they were adventurous or restless.

They went in trim, wooden-wheeled wagons pulled by oxen or mules. The wagons were called *schooners*, named after the fast two-masted ships that sailed out of New England's ports. These were prairie schooners, and they weren't fast. But, when the wind blew over the prairies and filled the canvas that roofed the wagons, they seemed almost to be sailing across the waves of prairie grass.

They called themselves *emigrants* because they were leaving the United States. The pioneers (for that is what we call them now) were going west to places they had heard about from wandering mountain men, or from traders, or from newspaper articles (many of which were written by people who had never left home).

They went due west. They were heading for foreign lands—Mexican

A mule pack train crossing the Bitterroot Mountains on a route that linked Fort Benton, on the Missouri River, and Fort Walla Walla, on the Columbia (in what is now Washington state). Even after a road was built, in 1862, pack trains took 35 days to travel the 624 miles.

California, or the territory of Oregon. (Oregon was claimed by England and the United States.) Most had no intention of settling in the U.S.–owned Louisiana Territory; they all knew what Stephen Long had said about that region.

It was a depression that got many of them started. A depression is a time of economic hardship when money seems to disappear and jobs do, too. In 1837, the United States entered a terrible depression. In New York and Philadelphia and Baltimore, thousands had no jobs. Banks closed and people lost their savings. The price of corn and wheat fell below the cost of growing it. Many farmers who had borrowed money to buy farming equipment and seeds couldn't pay back the loans, so they lost their farms.

What could they do? They could sell whatever they had left and use the money to head west (where land was free and fertile and opportunity seemed to be waiting). And, in 1843, that was just what some people began to do. They went west on the Oregon Trail.

Many of them never made it. They weren't prepared for the hard, hard journey: for rain that soaked through the canvas cover of the prairie schooner, for biting cold and burning heat, for hunger and accidents and disease. They brought the disease with them from the East. It was called cholera (KOL-er-uh), and it had come from Europe with the immigrants. It went west, as they did, and it was a killer. The trails became lined with graves; mothers and fathers buried their children—sometimes children buried their parents.

But they began the trip with optimism; they were looking for adventure.

CHOLERA.

THE

DUDLEY BOARD OF HEALTH,

HEREBY GIVE NOTICE, THAT IN CONSEQUENCE OF THE

Church-yards at Dudley

Being so full, no one who has died of the CHOLERA will be permitted to be buried after *SUNDAY* next, (To-morrow) in either of the Burial Grounds of *St. Thomas's*, or *St. Edmund's*, in this Town.

All Persons who die from CHOLERA, must for the future be buried in the Church-yard at Netherton.

BOARD of HEALTH, DUDLEY.
September 1st, 1832.

W. MAURICE, PRINTER, HIGH STREET, DUDL

Cholera epidemics were common until Americans built sewage systems and understood the importance of a clean water supply.

It is a 2,000-mile walk, over plains and mountains, from Missouri to Oregon. The pioneers set out in early spring and, if they make it on schedule, arrive before the winter snows.

Fort Laramie (in what is now eastern Wyoming) was founded by two of Ashley's mountain men. "Fort Laramie is a great place in the immigration season," wrote one man on his way to California. "A good many wagons are left at this point, many coming to the conclusion of getting along without them....A hotel, store, and post office are located here."

Deschutes, in French, means "falls" or "rapids." The Deschutes River was filled with rapids and was very dangerous.

Most were young—many of the mothers and fathers were still teenagers—and they added children as they went west. One in five of the women is said to have been pregnant on some part of the journey. Their babies were born in the hard-floored schooners, or in tents, or in the out-of-doors. Their children would remember the fun and the freedom of the trip. Martha Morrison went west to the Oregon Territory in 1844, when she was 13.

> *We did not know the dangers we were going through. The idea of my father was to get on the coast: no other place suited him, and he went right ahead until he got there.... We went down the river Deschutes in an open canoe, including all the children; and when we got down there was no way to get to the place where my father had determined to locate us, but to wade through the tremendous swamps. I knew some of the young men that were along laughed at us girls, my oldest sister and me, for holding up what dresses we had to keep from miring; but we did not think it was funny.*

For most pioneers, the journey west began in St. Louis. (The pioneer family might have already traveled from the East Coast on the Erie Canal, and then overland, and then down the Ohio River and on across the Mississippi.) St. Louis, where the Missouri River meets the

Mississippi, was now the gateway west. Pioneers who had money could take a steamboat from St. Louis, up the Missouri River, and head for a "jumping-off" town, like Independence, Missouri, or Council Bluffs, Iowa. There they could buy supplies and team up with other emigrants. Francis Parkman, who had come from Boston, described Independence in 1846:

> *The town was crowded. A multitude of shops had sprung up to furnish the emigrants and Santa Fe traders with necessaries for their journey; and there was an incessant hammering and banging from a dozen blacksmith's sheds, where the heavy wagons were being repaired, and the horses and oxen shod. The streets were thronged with men, horses, and mules. While I was in the town, a train of emigrant wagons from Illinois passed through, to join the camp on the prairie, and stopped in the principal street. A multitude of healthy children's faces were peeping out from under the covers of the wagons.*

Usually, only the smallest children and the sick got to ride in the wagons. There was no room for anyone else. The wagons weren't as big as the

O, Susanna,
Now don't you cry for me,
For I come from Alabama
With my banjo on my knee.
—WORDS FROM A POPULAR CAMPFIRE SONG WRITTEN IN 1848 BY STEPHEN FOSTER

In 1845, Francis Pettygrove (from Portland, Maine) and Asa Lovejoy (from Boston, Massachusetts) were in the Oregon territory laying out a town. They couldn't agree on a name, so they flipped a coin. Pettygrove won and named the settlement after his hometown in Maine.

St. Louis: Improving Considerably

Charles Dickens visited St. Louis in 1842. This was what he found:

In the old French portion of the town, the thoroughfares are narrow and crooked, and some of the houses are very quaint and picturesque: being built of wood, with tumble-down galleries before the windows approachable by stairs or rather ladders from the street. There are queer little barber's shops and drinking-houses, too, in this quarter: and abundance of crazy old tenements with blinking casements, such as may be seen in Flanders. Some of these ancient habitations, with high garret gable windows perking into the roofs, have a kind of French shrug about them; and being lop-sided with age, appear to hold their heads askew, besides as if they were grimacing in astonishment at the American Improvements.

It is hardly necessary to say, that these consist of wharfs and warehouses, and new buildings in all directions; and of a great many vast plans which are still "progressing." Already, however, some very good houses, broad streets, and marble-fronted shops, have gone so far ahead as to be in a state of completion; and the town bids fair in a few years to improve considerably though it is not ever likely to vie, in point of elegance or beauty, with Cincinnati.

Charles Dickens

George Catlin's painting of St. Louis and the river in 1832.

Nine miles out of Independence the wagon trains cross the Missouri state line and are in land guaranteed to the Indians (later the Indian treaties will be forgotten, and this will be Kansas). Thirty-two more miles and the road forks. A small hand-lettered sign points northwest and says, "Road to Oregon." Those who take the other branch are going southwest, on the Santa Fe Trail.

On the trail through Utah. One pioneer woman wrote: "Husband has had a tedious time with the wagon today. It got stuck in the creek this morning when crossing, and he was obliged to wade considerably in getting it out."

What is a chamberpot?

A Missouri farmer explaining why he was going to Oregon:
Out in Oregon I can get me a square mile of land. And a quarter section for each of you all. Dad burn me, I am done with the country. Winters its frost and snow to freeze a body; summers the overflow from Old Muddy [the Mississippi] drowns half my acres; taxes take the yield of them that's left. What say, Maw, it's God's Country.

heavy Conestogas the Pennsylvania Dutch used back East. The prairie schooners had to be lightweight so they wouldn't exhaust the oxen on the long pull ahead. Besides, there was so much to take that they were always filled.

Pretend you are leaving home—perhaps forever. What will you take with you? You need food for your trip: flour, beans, bacon, coffee, dried fruit, sugar, salt, and vinegar. Sheep, goats, cows, and chickens will come along. You will need clothes. Your mother packs pants, shirts, dresses, and some cloth to make clothes for you children as you grow. Into the wagon go pots, pans, water kegs, teakettles, and chamberpots to use along the way, and axes, plows, and saws, for the new life that is ahead of you. There are books: a Bible, schoolbooks, and storybooks. Your parents are musical: your father has brought his violin (he calls it a fiddle); your mother has packed a harmonica. In the evening, around the campfire, they entertain the others. There is still more in the wagon: a favorite family portrait, a mirror, and a rocking chair. There are guns,

medicines, and spare parts for wagon repairs. You have brought a hoop and some marbles; your sister has brought a doll; the baby, a rattle.

When the oxen are exhausted and lie dying by the trail, your parents will sell the plow, the axe, the books, and the teakettle, get two mules for them and be lucky. The cows will be gone—eaten when there was no game to be shot. The portrait and the rocking chair are gone also, left under a tree, perhaps for Indians to find. As you climb into the Rockies everything that adds unnecessary weight must be left behind—your lives may depend on that.

A Letter From Iowa

Some people were taking the Oregon Trail clear across the land, but others were settling in regions that bordered the Mississippi, such as Iowa and Kansas and Minnesota. Back East, everyone wanted to know: what were those far places like?

Well, it was frontier territory, and things weren't easy, but there was good land, and it was cheap and available. That was what mattered most. Jeremiah Fish traveled from New Jersey to the Iowa territory, and never regretted the move. This is what he wrote to the folks back home:

To: Mr. Samuel Rudderow
Pensaikin, near the city of Camden
New Jersey
From: Louisa County, Iowa Territory, April 25, 1843.

Brother and Sister,

It has been sometime since I wrote particularly to you, but you have heard from me frequent by letters to Rebecca Fish. I am in good health, but have had the hardest winter that I ever experienced. I got my feet badly frozen Dec. 8th and suffered more pain than I am able to describe, but my right foot has got well and I can wear my shoes but my toes are all stiff at the two outer joints and some of the bones came out of three toes. The toes of my left foot are all off and two of them healed over and the other 3 in a good way so that I will be but very little lame in a few more weeks....

When I wrote to you last I recommended this country to you and probably told you that it would be profitable for you to emigrate....I would be the last one to recommend such a thing to connections of mine if I was not positive, and I know NJ well, I understand Iowa well enough not to be mistaken, a man that pays rent or interest money in NJ had better be in Iowa, if he gets a farm paid for in a lifetime in NJ he had done well....here the land will produce at least 3 times as much to the acre, as it will there and with less labor....when I first came to Bloomington there was only 7 families there, now there is 15 good stores there and merchandise of all kinds and the cost has improved as fast as the town, but here with only $80 you would be better off than you would be there in 200 years, even if you could live so long. We do not tell you this under any fake pretense but for the benefit of you and family, weigh the matter and write me and I will tell you more.

Respectfully yours,
Jeremiah Fish

"Another Fine Cow Died This Afternoon"

Amelia Stewart was born in Boston. That was where she met Joel Knight, who had come from England and was studying to be a doctor. They got married and a few years later headed west, to Iowa. In 1853, after 16 years in Iowa, they packed their goods in a covered wagon and headed west again. By now they had seven children. Amelia Knight kept a diary of their five-month journey. The actual diary, in her handwriting, can be found at the University of Washington library. Here is some of it.

✍ WEDNESDAY, JULY 27TH—Another fine cow died this afternoon. Came 15 miles today, and have camped at the boiling springs, a great curiosity. They bubble up out of the earth boiling hot. I have only to pour water on my tea and it is made. There is no cold water in this part....

✍ SUNDAY, JULY 31—Cool and pleasant, but very dusty. Came 12 miles and camped about one o'clock not very far from Boise River....

✍ MONDAY, AUGUST 1ST—Still in camp, have been washing all day, and all hands have had all the wild currants we could eat. They grow in great abundance along this river. There are three kinds, red, black, and yellow. This evening another of our best milk cows died. Cattle are dying off very fast all along this road.

"This is a land of wonders and of hardships; a land to be avoided or left behind as soon as possible. Saw many dead cattle on the road; the poisonous water and the great scarcity of feed begin to tell on the poor brutes."

We are hardly ever out of sight of dead cattle on this side of Snake River. This cow was well and fat an hour before she died. Cut the second cheese today....

✍ FRIDAY, AUGUST 5TH—We have just bid the beautiful Boise River, with her green timber and rich currants farewell, and are now on our way to the ferry on the Snake River. Evening—traveled 18 miles today and have just reached Fort Boise and camped. Our turn will come to cross some time tomorrow. There is one small ferry boat running here, owned by the Hudson's Bay Company. Have to pay three dollars a wagon. Our worst trouble at these large rivers is swimming the stock over. Often after swimming half way over the poor things will turn and come out again. At this place, however, there are Indians who swim the river from morning till night. There is many a drove of cattle that could not be got over without their help. By paying them a small sum, they will take a horse by the bridle or halter and swim over with him. The rest of the horses all follow and by driving and hurrahing to the cattle they will almost always follow the horses, sometimes they fail and turn back.

✍ MONDAY, AUGUST 8TH—We have to make a drive of 22 miles, without water today. Have our cans filled to drink. Here we left, unknowingly, our Lucy behind, not a soul had missed her until we had gone some miles, when we stopped a while to rest the cattle; just then another train

A pioneer daughter recalled: "When our covered wagon drew up beside the door of the one-roomed sod house that father had provided, he helped mother down and I remember how her face looked as she gazed about that barren farm, then threw her arms about his neck and gave way to the only fit of weeping I ever remember seeing her indulge in."

drove up behind us, with Lucy. She was terribly frightened and said she was sitting under the bank of the river, when we started, busy watching some wagons cross, and did not know we were ready....It was a lesson for all of us.

✍ FRIDAY, AUGUST 12TH—Lost one of our oxen. We were traveling slowly along, when he dropped dead in the yoke....I could hardly help shedding tears, when we drove round this poor ox who had helped us along thus far, and had given us his very last step.

✍ THURSDAY, AUGUST 18—Commenced the ascent of the Blue Mountains. It is a lovely morning and all hands seem to be delighted with the prospect of being so near the timber again, after the weary months of travel on the dry, dusty sage plains, with nothing to relieve the eye. Just now the men are hallooing till their echo rings through the woods. Evening—Traveled 10 miles today and down steep hills, and have just camped on the banks of Grand Ronde River in a dense forest of pine timber—a most beautiful country.

✍ FRIDAY, AUGUST 19TH—Quite cold morning, water frozen in the buckets. Traveled 13 miles, over very bad roads, without water. After looking in vain for water, we were about to give up as it was near night, when husband came across a company of friendly Cayuse Indians about to camp, who showed him where to find water, half mile down a steep mountain, and we have all camped together with plenty of pine timber all around us....we bought a few potatoes from an Indian, which will be a treat for our supper.

The last entry in Amelia's diary came on SATURDAY, SEPTEMBER 17; *they were in Oregon.*

✍ A few days later, my eighth child was born. *[Never, in the diary, had she mentioned being pregnant.]* After this we picked up and ferried across the Columbia River, utilizing skiff, canoes and flatboat to get across, taking three days to complete. Here husband traded two yoke of oxen for a half section of land with one-half acre planted to potatoes and a small log cabin and lean-to with no windows. This is the journey's end.

6 Getting There

A Woman Alone

Janette Riker was on her way to Oregon with her father and two brothers. They were in Montana—it was September 1849—and the men went off hunting. They never returned. Riker was on her own. She knew she couldn't cross the mountains alone, and winter was coming, so she took tools from the wagon and built a hut for herself. Then she put the wagon's stove inside, along with provisions, and blankets, and a load of firewood she had chopped. Because she had no meat, she killed an ox, butchered and salted it, and prepared for Montana's long, cold winter. It soon came. Wolves and mountain lions sniffed outside her shelter, but she didn't budge from the hut until a spring thaw flooded her out. Then she moved back into the wagon. Finally, Indians found her and were so astonished that she had survived—on her own—that they took her west, where she wanted to go.

Settlers in the northwest in 1825. Guns take pride of place on the walls, along with two or three books (probably a Bible and prayer book)—and a coffee grinder.

The pioneers didn't go west alone. That would have been foolish. The journey was too dangerous. They traveled in groups of wagons, and they called them wagon trains.

They met in Independence, or one of the other jumping-off towns. Sometimes they didn't even know each other when they began the trip. Before long, the wagon train seemed like a big family. Often the pioneers made plans to settle together. Just as the people who traveled on the *Mayflower* made a compact that gave them rules and leaders, so too the wagon communities wrote their own constitutions and had laws, courts, and officers.

Jesse Applegate's group did just that. He was one of the leaders of the first big wagon train to Oregon. It must have looked like an enormous parade—with wagons, cattle, chickens, dogs—and more than 1,000 pioneers. (About 600 of them were children.) It was a well-organized trip. Baptiste Charbonneau was a guide. Remember Charbonneau? He was Pompey, Sacajawea's baby, now grown-up. Here is Jesse Applegate describing a day on the trail:

> *It is four A.M.; the sentinels on duty have discharged their rifles—the signal that the hours of sleep are over; and every wagon and tent is pouring forth its night tenants, and slow-kindling smokes begin to*

rise....From six to seven o'clock is a busy time; breakfast is eaten, the tents struck, the wagons loaded, and the teams yoked. There are 60 wagons. They have been divided into 15 divisions or platoons of four wagons each. The women and children have taken their places [in the wagons]. The pilot stands ready to mount and lead the way. Ten or fifteen young men are ready to start on a buffalo hunt....

It is on the stroke of seven that the clear notes of the trumpet sound in the front; the leading division of wagons moves out of the encampment and the rest fall into their places...the wagons form a line three quarters of a mile in length; some of the teamsters ride upon the front of their wagons, some walk beside their teams; scattered along the line companies of women and children are taking exercise on foot; they gather bouquets of rare and beautiful flowers that line the way.

The pioneers walk all day with a stop for lunch. As the sun is setting, the wagons halt and cluster together.

It is not yet eight o'clock when the first watch is to be set; the evening meal is just over....near the river a violin makes lively music, and some youths and maidens have improvised a dance; in another quarter a flute gives its mellow and melancholy notes to the still air....It has been a prosperous day; more than twenty miles have been accomplished.

They Took the Laws Along, Too

Pioneer Lansford W. Hastings set out for Oregon on May 16, 1842, as part of a wagon train of emigrants. They began in a spirit of "high glee" and "hilarity," he wrote in his journal.

We had proceeded only a few days travel, from our native land of order and security, when the "American character" was fully exhibited. All appeared to be determined to govern, but not to be governed. Here we were, without law, without order, and without restraint; in a state of nature.

Each emigrant, said Hastings, had his or her own ideas about how the journey should proceed. Hastings described them:

Some were sad, while others were merry; and while the brave doubted, the timid trembled! Amid this confusion, it was suggested by our captain, that we "call a halt," and pitch our tents, for the purpose of enacting a code of laws, for the future government of the company.

And that is exactly what they did. They made laws, and that kept order in the wagon train. Americans took their laws with them, into their wilderness settlements.

Eastward I go only by force, but westward I go free.

—Henry David Thoreau

***Nooning on the Platte*, by a famous painter, Albert Bierstadt. This midday rest time was very important for the animals.**

South Pass is the Wyoming mountain gap that Jedediah Smith found. If you went through the South Pass you didn't have to climb over the high Rockies.

Twenty miles, a fine day! Usually a wagon train does well to make 15 miles a day. How about doing some arithmetic? It is about 2,000 miles from Missouri to the West Coast. If you walk 2,000 miles and you average 15 miles a day, how many days will it take to get to your destination? Now turn those days into months.

After you get that answer you can add a month: for river crossings, wagon repairs, a snowstorm, a much-needed rest, and a day for the birth of a baby. If those are your only delays you will be very lucky.

You have a choice of trails west. Jesse Applegate's wagon train went down the treacherous Snake River, then across 300 miles of desert, and then climbed the Blue Mountains into Oregon. You're going to take a different route. Let's pretend that you are at South Pass and on your way to California. You are 10 years old and in the midst of an adventure you will never forget. You like the pattern of the days. You like floating across rivers in a wagon, outdoor living, eating buffalo steaks, drinking water from clear mountain streams, and having other children to play with. You do not like the mosquitoes, the pounding rainstorms, buffalo hump soup, the day there is no water to drink, the sadness when a horse stumbles and your best friend's father is killed. But, all in all, it has been a good trip—so far.

Below, the daunting scenery facing the pioneers: the Sierra Nevada mountains and (bottom) the Colorado Desert looking toward Signal Mountain. "O for a lodge in some vast wilderness," wrote one young woman, "a home in some deep lone 'kenyon.'"

At South Pass you are halfway to California; there is excitement in the wagon train. The air is clear and the climb is gradual. The pass is like a roadway between mountains. At your feet are yellow violets, purple larkspur, and a few stalks of that red flower called Indian paintbrush. The birds here are as colorful as the wildflowers: you spot mountain bluebirds and yellow tanagers. High overhead a bald eagle circles. It is August, but you can see snow and ice on top of the mountains. You don't realize how high above sea level you are. You also don't realize what is ahead of you. Your parents don't either.

The pass is full of game: pronghorn, elk, bison, and bighorn sheep with great curving horns. Your father kills an elk and everyone celebrates and feasts around the campfire.

You come down from the Rockies in high spirits. The Utah desert discourages you a bit, but the Humboldt River with its grassy banks cheers you again.

Soon the Humboldt turns foul. People and animals who drink it become sick. Then the river ends. Just plain ends. There is nothing but desert and burning

sun. No one has ever heard of a river just ending. The Humboldt does.

Now there are 65 miles to walk with no water at all. Sixty-five miles littered with skeletons of mules, oxen, and people.

Do you wish you hadn't come? Some people go crazy—really crazy; they lose their minds under the beating sun of the desert.

It isn't over when you reach the Truckee River. The mountains ahead will make the Rockies seem easy, even though they are not as tall. The Sierra Nevada are real rocky mountains. Made of granite, they are, and, when the snows come, almost impossible to climb.

Of course, you can avoid them. You can go down to Death Valley. But I wouldn't suggest that. John Bidwell made it over the Sierras. It won't be easy, but you can make it too.

Bidwell was a young schoolteacher who went off on a vacation. When he came back to his home in Missouri, someone had stolen his land. That someone was bigger and had a gun, so Bidwell decided not

The California trail follows the Humboldt River for about 350 miles—much of it pleasant, with water, grass, and game for food. Then the Humboldt sinks beneath the desert in an area called the Humboldt Sink. From there to the Truckee River there is no water.

In 1848 pioneers found another route to California. It went along the Carson River to Lake Tahoe and was a bit easier than the Truckee Route, 25 miles to the north.

One journey's end: Oregon City on the Willamette River.

Traveling in a Palace

Virginia Reed headed west in style. It was 1846, she was 13, and her family traveled in something they called a "pioneer palace car." The "car" had a roomy second story used as a bedroom. Downstairs was a living room with high-back chairs and an iron stove that warmed the wagon. A mirror hung opposite the entrance; Virginia said, "knowing that books were always scarce in a new country, we also took a good library of standard works." Virginia made it to California, but the palace car and the rest of her family didn't. They teamed up with some people named Donner. But that's a story you'll have to read about on your own.

to fight. Besides, he met a Frenchman who had been to California and told him that it was always springtime there and that no one ever got sick. (Yes, Americans do go for tall tales.)

Bidwell decided to walk to California with wagons hauling his belongings. He talked others into going with him. They thought they could just head west—into the setting sun—and they'd find California. Luckily, they met Tom Fitzpatrick, who had been one of Ashley's mountain men. Fitzpatrick was on his way to Oregon, so he guided them much of the way. Then Fitzpatrick went north. Bidwell and his party took a cutoff that went southwest.

Their wagons never got over the Sierra Nevadas, nor their oxen. They ate the oxen to keep from starving. Then Bidwell and his friends kept walking. They made it. They got to California. It was 1841.

In 1844 the first wagons made it all the way—pulled up one side of the steep mountains with ropes and pulleys and let down on the other side the same way. After that the pioneers started flowing west, like a river.

In 1845, 3,000 pioneers traveled west on the Oregon and California trails. During the following two years, more than 5,000 religious pioneers left the United States and headed for the valley of the Great Salt Lake. They set up communities in a land they called Deseret. It was soon to be known by its Indian name: *Utah.*

Remember how the Pilgrims came to America so they could practice their religion in peace? Well, 200 years later, these religious pioneers had to leave the United States for the very same reason.

A general store in Omaha City, Nebraska. The wagons' destinations are written on their covers.

7 Latter-Day Saints

Joseph Smith was in jail in Illinois on treason charges when a mob dragged him out and shot him.

Like the Pilgrims and Puritans, they called themselves "saints." Like the early Quakers, they were mostly poor folk. Like those other believers, they were willing to suffer persecution for beliefs that differed from those of the mainstream churches. They were Mormons, members of the Church of Jesus Christ of Latter-day Saints.

Theirs was a new religion, a made-in-America religion, a Bible-based religion founded by Vermont-born Joseph Smith, who had a vision of "two personages [God and Jesus], whose brightness and glory defy all description." Smith said he was led by an angel to find a holy book, the Book of Mormon, engraved on golden tablets. It told the story of an ancient Hebrew tribe that came to America in long-ago times and to whom Jesus had appeared.

Smith, who was raised in Palmyra, New York, founded his religion in 1830 with six followers; within a few years there were thousands of Mormons. That name, *Mormon*, was not one they chose for themselves. Just as the words *Puritan* and *Quaker* had been chosen by others, so, too, with *Mormon*. The name of their church was too long for most people to bother saying. "Mormons," after their holy book, was an easy label. After a while it was used proudly by those who saw themselves as saints.

The Mormons were determined to follow their religion. They formed a tight community, a church-state where they shared their goods, worked hard, and listened to strong leaders. There were those who tried to stop them.

Why would people attack devoted churchgoers?

Some people may have been jealous of the success of the Mormons;

Like the Mormons, the Shakers built a religious community where people put aside personal goals for a common ideal. People called those communities *utopian* (yoo-TOE-pee-un), because they tried to be perfect. There were other utopian experiments in the 19th century, but eventually they all failed—except for the Mormons.

The land the Mormons chose—in what is now Utah—is known to geographers as the Great Basin: it is a plateau lying between the Wasatch and Sierra Nevada ranges. No other pioneers seemed to want it.

During its heyday, the Mormon city Nauvoo, on the Mississippi, had between 15,000 and 20,000 inhabitants. Today you may visit Joseph Smith's original log cabin; the town's population is now 1,100.

Brigham Young

and some people fear those who seem different. The early Mormons practiced *polygamy* (puh-LIG-uh-mee). Polygamists have more than one wife. Founder Joseph Smith had a whole lot of wives. Polygamy upset many non-Mormons, who thought it morally wrong. (There were more women than men among the early Mormons. Polygamy may have solved that problem.) Mormons tried to convert others to their religion. That made some people very angry, especially those whose children got converted.

The First Amendment to the Constitution forbids religious persecution. That didn't stop some people. Mormons were persecuted. First they moved to Ohio, then to Missouri, and then to Illinois. In Illinois they built a beautiful city—called Nauvoo—with a temple and a university. It was the largest city in Illinois, larger than Chicago. But even in Nauvoo, the Mormons weren't safe from the religious bigots: mobs attacked, destroyed much of the city, and murdered Joseph Smith.

Brigham Young was campaigning for Joseph Smith for president when he heard of the murder. He rushed back to Nauvoo and became the new Mormon leader. (Some Mormons followed Joseph Smith's son, and they formed a separate Mormon church.) Young was a strong, inspiring man with a lot of common sense. (He had two dozen wives

In 1844, Warsaw, Illinois, passed a resolution: "The adherents of Smith, as a body, should be driven...into Nauvoo...a war of extermination should be waged, to the entire destruction...of his adherents." Two years later, the Mormons were expelled from Nauvoo itself.

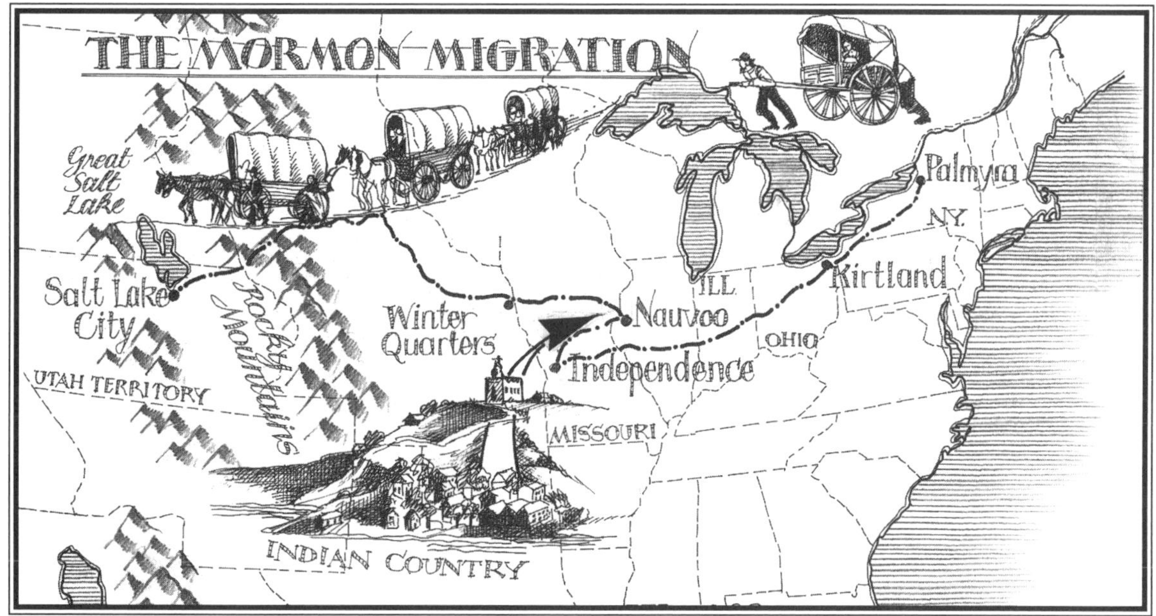

and 58 children. You can decide if that made sense or not.)

The next two years were violent. There were more attacks on these religious believers; but now the Mormons fought back with revenge raids. Finally, Young decided to lead his people west, out of the United States, to land held by Mexico. It was land where Mormons could work hard, be productive, and be left alone to follow their beliefs (land only the mountain men and the Indians knew).

On their way to Utah, the Mormons—10,000 of them—spent the winter of 1846–1847 in temporary quarters in what is now Omaha, Nebraska.

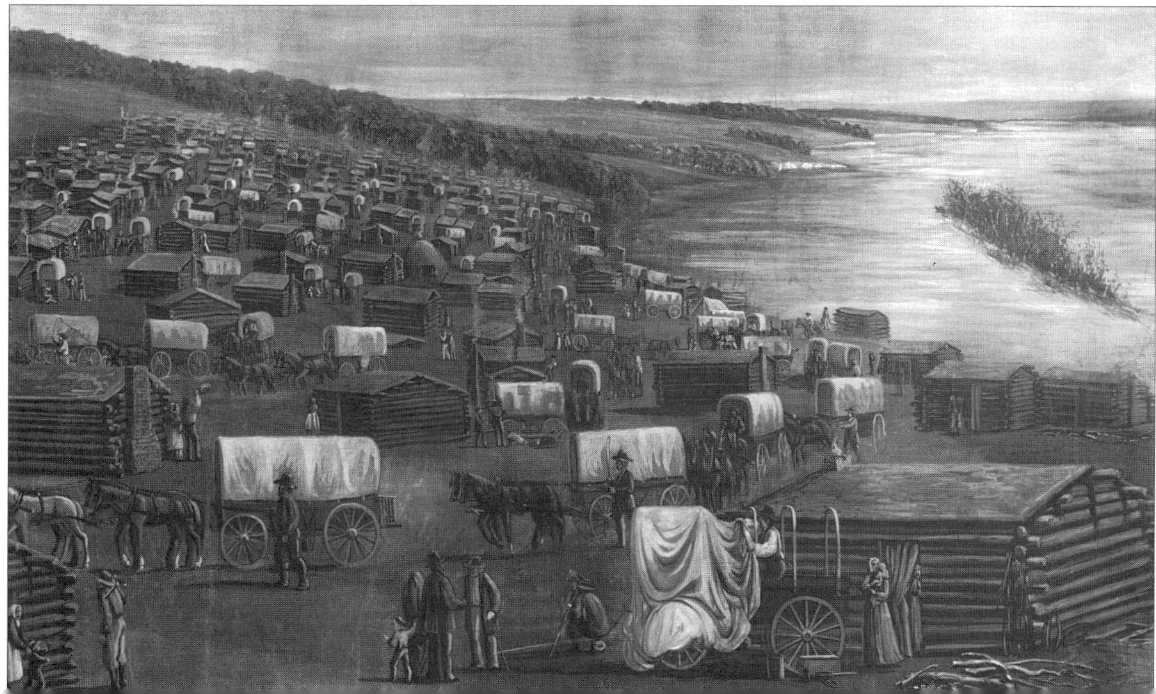

This was the desert facing travelers west of the Great Salt Lake. If it hadn't been for the Mormons' help at the start of this leg of the trip, others would have died trying to cross it.

"Our pioneers are instructed to proceed west until they find a good place to make a crop...." Three years after the Mormons' migration, Salt Lake City was thriving.

In 1845, Young sent a few Mormons west to check out the Great Salt Lake. They liked what they saw. The next winter was shiveringly cold, much colder than usual, and the mighty Mississippi froze into a solid roadway. Brigham Young and the other Mormons crossed the great river on foot and on horseback. They left Nauvoo behind; it was a city of ghostly memories. By the summer of 1847, a long ribbon of Mormon wagon trains was moving across the dusty trail west. An old trail song gives you an idea of what it was like:

They came to the desert and salt water lakes
The ground it was teemin' with varmints and snakes.
Beset by wild Indians, Comanche and Sioux,
'Tis a glorious tale how they ever got through!

It wasn't easy crossing the treeless plains, or climbing into the mountains. Many died doing it. Many Mormons died. But Brigham Young planned well. His may have been the best organized of all the western treks. Certainly it was the largest. Young sent groups ahead to create stopping places, build shelters, plant crops, mark the trail, prepare for the wagon trains.

And so they reached the Great Salt Lake and built

The Mormons who went with Brigham Young brought a boat along with them. They used it to cross rivers. When they reached the North Platte River, near Laramie, Wyoming, they left the boat behind. There it stayed for 20 years, taking pioneers (for a fee) across the river. It was known as the Mormon Ferry.

their religious kingdom and made desert land bloom. When Brigham Young died, in 1877, there were 140,000 Mormons living in Utah in 325 towns. They had built railroads, stores, factories, and irrigation projects. Mormon troubles weren't over, but they soon would be. Mormons would give up polygamy, the Utah territory would become a state, and Mormons would become an accepted part of the pluralism—the many-sidedness—that makes our country strong and interesting.

But before that, just as the Mormons were beginning to build their community beside the Great Salt Lake, their settlement became very important. Suddenly a stream of pioneers was heading west. They were people on their way to Oregon and California. Many would have starved and died if Mormons hadn't supplied food, horses, and a place to rest. In 1849, the tide of pioneers turned into a flood. That year 55,000 people followed the Overland Trail, most of them taking the cutoff that led southwest, to California. They were very anxious to get to California. (You'll find one reason for that in the next chapter; the big reason comes later.)

Brigham Young urged poor Europeans to come to America and become Mormons. In 1856, 2,500 of them walked the 1,300 miles from Iowa City to Utah pulling their possessions in handcarts (with no help from mules or oxen). Other Mormons, before and after, came in wagons.

This satirical cartoon—making fun of polygamy—was published after Brigham Young's death in 1877.

8 Coast-to-Coast Destiny

"The world beholds the peaceful triumphs of ...our emigrants," said President Polk. "To us belongs the duty of protecting them...wherever they may be upon our soil."

President James K. Polk wanted California and Oregon, and so did most other Americans. The land was enticing, and there was something that convinced people that it was right to take it. It was an idea called *manifest destiny*. Those highfalutin words were first used by a newspaper reporter. He said it was the manifest destiny of the United States to fill the land from coast to coast. Polk believed that America had the right and duty to spread democracy across the continent. Most Americans agreed with him. Soon that phrase *manifest destiny* was on everyone's tongue.

Many Californian Indians, such as these men in war-dance costumes from San Francisco, had to work on Mexican missions. But the coming of the U.S. meant worse times for them. By the year 1900 more than 80 percent of the tribes had vanished.

Both Great Britain and the United States claimed Oregon. In 1846, when President Polk signed a treaty with England, it settled the Oregon problem. The two nations agreed to split the Oregon territory on the 49th parallel. That was the 49th line of latitude, counting up from the equator. England got the land north of the parallel; the United States, the land to the south. (That U.S. land today is Oregon, Washington, Idaho, and parts of Montana and

Nothing less than a continent can suffice as the basis and foundation for that nation in whose destiny is involved the destiny of mankind. Let us build broad and wide those foundations: let them abut only on the everlasting seas.

—IGNATIUS DONNELLY, 19TH-CENTURY POLITICIAN AND REFORMER

Wyoming. The English land is western Canada.)

California belonged to Mexico. People in the United States didn't know much about the West Coast. Then Richard Henry Dana, who had been to California on a sailing ship, wrote about his adventures in a book called *Two Years Before the Mast*. It was published in 1840 and was good reading. President Polk and other Americans became excited about California.

Richard Henry Dana

Dana had been surprised by the Californians: by their good manners and their fine clothes. They spoke Spanish, they were as elegant as old Spain, and they lived with the ease that seems to come naturally in a sunny climate.

> *Next to the love of dress, I was most struck with the fineness of the voices....Every common ruffian-looking fellow...appeared to me to be speaking elegant Spanish. It was a pleasure, simply to listen to the sound of the language....A common bullock-driver, on horseback, delivering a message, seemed to speak like an ambassador at a royal audience.*

The Californians didn't appear to have money, as people did in the east. Instead they used silver or cattle hides. *I certainly never saw...so much silver at one time, as during the week that we were at Monterey,* wrote Dana. *The hides they bring down dried and doubled, in clumsy oxcarts or upon mules' backs.* Sailors called the hides "California bank notes."

But what about the Native Americans? Dana made it clear that they had lost control of their land. How did it happen?

Manifest Destiny

The phrase *manifest destiny* was first used by John L. O'Sullivan in an article published in 1845 in the *United States Magazine and Democratic Review.* O'Sullivan said it was "our manifest destiny to overspread the continent allotted by Providence for the free development of our yearly multiplying millions." No one wanted to admit that greed might have something to do with the drive to take more land. Manifest destiny made it sound like a noble thing to do. In this picture a goddess of progress, towing a telegraph wire, shepherds the settlers on their way west. The Indians, bison, and bear have been pushed off the western edge into extinction—which proved to be all too true.

Indians had lived in California for thousands of years, cut off from the rest of the world. Then, after Magellan found a way to sail around South America, explorers and ships made occasional visits. Juan Cabrillo went

Spanish Mission
Indian Quarters
Corral
Stable
Black-smith
Carpentry
Women's Quarters
Weaving and Carding
Storage
Soap and Tallow Vats
Tanning Vats
Garden
Soldiers' Quarters
Cemetery
Guests' Rooms
Priests' Quarters
Church
INDIANS AT WORK STAKING OUT A HIDE AND BRINGING ROCKS TO BUILDING SITE.

there and so did Francis Drake. The explorers didn't stay, but they did leave their germs—and those germs killed a lot of Indians. Most of the California natives were wiped out. In 1769 (about the time Tom Paine and Patrick Henry were stirring things up on the other side of the continent), the Spanish government sent a few priests and soldiers to California. Spain didn't really seem interested in the place, but it looked as if the Russians might be thinking about moving in, and Spain didn't like that idea.

The priests built Catholic missions along the coast. The missions were agricultural settlements—with lovely gardens and churches—where Indians were taught about Christianity and made to do all the hard work of farming and building. The priests believed they were serving God by baptizing and teaching the Native Americans. They planned, one day, to give the missions to the Indians. Some of those missions were very prosperous. The Mission San Gabriel had 150,000 cattle, 20,000 horses, and 40,000 sheep, along with orchards, a distillery, looms, and more.

Some of the Spanish settlers were convicts sent to California to get them out of Spain or Mexico (the English did the same thing and sent many of their prisoners to their colonies).

When Mexico won its independence from Spain, in 1821, Mexicans became rulers of California. Spain no longer controlled the missions. The mission land, which had been intended for the Indians, was taken by Mexican Californians, who turned most of it into huge cattle ranches. The Native Americans became the workers on those ranches; they were almost like slaves. "The Indians," Dana wrote, "do all the hard work." But for the *rancheros*—the owners of the big ranches—life seemed mighty good. Dana described it:

In 1846, some Mexican rebels take over the city of Sonoma and hoist a flag with a bear on it. They say that they are independent of Mexico, and they call their nation the republic of California. When American John Frémont is made leader of the republic, the Stars and Stripes replaces the Bear flag and Mexicans and Americans begin fighting. (More on that soon.)

> *Horses are as abundant here as dogs and chickens [elsewhere]. There are no stables to keep them in, but they are allowed to run wild and graze wherever they please, being branded, and having long leather ropes, called lassos, attached to their necks.... The men usually catch one in the morning, throw a saddle and bridle upon him, and use him for the day, and let him go at night, catching another the next day. When they go on long journeys, they ride one horse down, and catch another, throw the saddle and bridle upon him, and after riding him down, take a third, and so on to the end of the journey. There are probably no better riders in the world.*

The British naval officer and official artist who painted this picture called it the "California Mode of Catching Cattle."

There is a power in this nation greater than either the North or the South—a growing, increasing, swelling power, that will be able to speak the law to this nation, and to execute the law as spoken. That power is the country known as the Great West.... There is the hope of this nation.

—SENATOR STEPHEN A. DOUGLAS

If you visit the city of Sacramento you can see Sutter's fort looking much as it did in the 1840s.

John Augustus Sutter's name was soon to be known all over the world. You'll find out why in chapter 11.

Since Mexico was having problems at home (it had 30 presidents in its first 50 years of independence), California was left on its own, which was the way most Spanish-speaking Californians wanted it anyway. There weren't many of them—perhaps 6,000 by 1840. No one knows how many Indians there were, but it was a lot more than that. Of others—Europeans (including Russians), Americans, Asians—there were about 400 in all.

One of the most interesting of the Europeans was a Swiss: John Augustus Sutter. He had become a Mexican citizen, acquired a huge land grant, and built a fort-plantation inside walls of adobe. Sutter's fort was like a little kingdom, where crops were grown, clothing made, horses raised, and soldiers trained. Sutter was unusually generous: to visitors, to the Indians who worked for him, to almost everyone. If you needed a meal, or a bed for the night, or some help or friendly advice—you could count on John Sutter. Many people did.

Especially those who began coming overland from another country—the United States—to settle. Richard Henry Dana thought they were just what California needed. To Dana, who was a serious-minded New England Yankee, the Californians seemed lazy. It was only "the character of the people," he said, that prevented California from becoming great.

> *The soil is as rich as man could wish; climate as good as any in the world; water abundant, and situation extremely beautiful....In the hands of an enterprising people, what a country this might be!*

Of course, you know whom he had in mind when he mentioned enterprising people. Sure you do—the same people President Polk had in his mind: Americans.

Between 1841 and 1848, a few thousand Americans made it over the mountains to the California valleys. They were enterprising people. But not everyone was happy to see them. "We find ourselves suddenly threatened by hordes of Yankee immigrants," the Mexican governor wrote. "They are cultivating farms, establishing vineyards, erecting mills, sawing up lumber, building workshops, and doing a thousand things that seem natural to them." Remember, when they arrived in California those Americans were foreigners in Mexican territory.

The land they found was as nature had made it—spectacular. Near the coast, fields of yellow-orange poppies stretched farther than anyone could see. In the mountains, waterfalls dropped a third of a mile and more, and trees grew taller than a 20-story building (although no building had yet been built that high). Prickly pear, primrose, and yucca blossomed in the desert.

As I said, President Polk wanted California. And California, at the time, meant more than today's West Coast state. It meant the land controlled by Mexico that included all or parts of the future states of New Mexico, Arizona, Utah, Nevada, Colorado and Wyoming, as well as California. President Polk sent a diplomat to Mexico City with an offer to buy that land. The Mexicans were insulted. They said they wouldn't sell, and certainly not to the United States; not after what the United States had just done in Texas. You'll read about that in the next chapter.

What John L. O'Sullivan did in print for manifest destiny, Emanuel Leutze (who made a famous painting of George Washington crossing the Delaware in the Revolutionary War) did on canvas with this picture, called *Westward the Course of Empire Takes Its Way*. Some rugged-looking mountain men guide pioneer families on their journey west.

9 Texas: Tempting and Beautiful

Stephen Austin wanted to encourage American settlers in Texas, so he got a law passed that allowed people to bring in slaves.

Texas was tempting. All that beautiful land...land that seemed made just to grow cotton.

That Texas land was part of Mexico and controlled by Spain. It had been so since the 16th century, when Spanish explorers—de Soto, Cabeza de Vaca, Fray Marcos, Estebán, and Coronado—had searched for the seven cities of Cíbola.

Those seven cities had turned out to be imaginary. But the invaders carried some things with them that were real—although no one then could even imagine them. They were germs; germs that killed most of the Native American population. So, in the early 19th century, the vast land of Texas was almost empty of people. Perhaps 30,000 Indians lived there, and a few thousand Spanish Mexicans. (Thirty thousand may sound like a lot of people, but not in Texas. A 20th-century sports stadium—the Houston Astrodome—holds 54,816 people.) The Spaniards had built missions in Texas where priests lived, farmed, and attempted to convert Indians to Christianity. Soldiers lived in *presidios*, which is the Spanish word for forts. The presidios protected the missions. Some Mexicans were ranchers and lived on *haciendas*, which were ranch plantations where cattle and crops were raised. Spanish-speaking cowboys were called *vaqueros*.

The name Texas comes from an Indian word, *texía*, for "friends" or "allies." It was the word a group of Indian tribes used to describe themselves. The Spanish turned it into *tejas* (TAY-huss), and the Anglos (English-speakers) said Texas.

Spain had hoped that some of its citizens would settle in Texas—as European settlers settled in the United States—but not many did. There was no gold in Texas, and no political or religious freedom. In the Spanish colonies everyone was expected to be Catholic.

That didn't bother Stephen Austin. In 1821 he led a group of 300 settlers to Texas from Missouri. They said they would become good

Mexican citizens and Catholics. That same year, Mexico rebelled against Spanish rule and became independent. Three years later, in 1824, the Mexicans approved a fine constitution and formed a republic. Unfortunately, there was no tradition of self-government in the Spanish colonies (as there had been in the English colonies). The people weren't used to running things themselves. That made it easy for strong, ambitious people to take power, and soon a dictator named Antonio López de Santa Anna took over. Santa Anna ignored many of the freedoms the constitution had promised.

Other people from the United States began settling in Texas. Some of them didn't want to be Mexican citizens. Some of them didn't want to become Catholics. They wanted schools and freedom of religion. They wanted to build towns and to run those towns themselves. Some brought slaves—which was against Mexican law. Some didn't want to share the land with Indians. Some even bragged about killing Indians.

By 1830, there were more English-speaking Americans in Texas than Mexicans. Santa Anna said no more Americans would be allowed to settle in Texas. That didn't stop them. People from the States crossed the border illegally and settled in Mexican territory. They demanded the rights of the Mexican Constitution of 1824. The issues and conflicts were complicated. You can see there would soon be trouble.

In 1836, Santa Anna marched toward Texas with an army of more

The cattle that *vaqueros* (above, with their horses) tended weren't bred for beef; until railroad cars were refrigerated no meat could be sent long distances. So their cattle were used for tallow (fat for candles) and for their hides—leather.

General Antonio López de Santa Anna saw himself as "the Napoleon of the West."

than 3,000 men (by the time they got to Texas the army was even larger). He was determined to do something about the *anglos* (which is what he called the English-speaking Americans). Some of the English-speakers went back to the United States, but some decided to stay and fight. The fighters gathered in the chapel of an old, walled San Antonio mission named the Alamo. There weren't even 200 of them, but they included some famous frontiersmen. Davy Crockett was there; so was Jim Bowie.

An ox cart outside San Antonio in old Mexican Texas.

Alamo means *poplar* in Spanish. A grove of poplar trees grew near the mission.

Crockett had a rifle he called "Old Betsy." He said Old Betsy had killed 105 bears in one season, and maybe that was true. Crockett was a great storyteller. He said it was his storytelling ability, not his speech-making skills, that got him elected when he decided to go into politics.

The thought of having to make a speech made my knees feel mighty weak, and set my heart to fluttering almost as bad as my first love-scrape. But as good luck would have it, those big candidates spoke nearly all day, and when they quit, the people were worn out with fatigue, which afforded me a good apology for not discussing the government. But I listened mighty close to them and was learning pretty fast about political matters. When they were all done, I got up and told some laughable story and quit. I found I was safe in those parts, and so I went home and I didn't go back again until after the election was over. But to cut this matter short, I was elected, doubling my competitor, and nine votes over.

Davy Crockett

Giving a campaign speech may have scared Davy Crockett, but once he got to the United States Congress he was fearless. He stood up and spoke out—even when he disagreed with the president, popular Andrew Jackson. When President Jackson sent the Cherokee Indians from their homes in Georgia to the Oklahoma territory, Davy Crockett thought it was wrong and said so. It didn't help the Indians, or Crockett. He lost his seat in Congress. But he kept, he said, "my conscience and my judgment." Crockett's autobiography made him famous. It told of his adventures as a woodsman and trailblazer.

March 5: Pop, pop, pop! Boom, boom, boom! throughout the day. No time for memorandums now. Go ahead! Liberty and independence forever!

—Davy Crockett

Jim Bowie was a fearsome fighter who designed a wicked, razor-

Jim Bowie

sharp knife that is still called by his name. Crockett and Bowie and the others in the Alamo held out for 12 days against Santa Anna's army. They had no reinforcements and not enough ammunition. Finally, the Mexicans broke through the walls of the Alamo and killed them all—except for three women and three children who lived to tell the story.

The remaining English-speaking settlers in Texas were furious. They decided to fight for independence for Texas. When they did, they yelled "Remember the Alamo!"

Their leader was Sam Houston. Sam was unusual, mighty unusual. He had been born in Virginia but, when he was 13, his parents moved to the frontier of Tennessee. At age 15 he was working in a store in Tennessee when he ran away from home. He became a Cherokee Indian—adopted into a tribe—and lived with the Indians for three years. Then he came back to the white settlements. These are some of the things he did: taught school, fought in Andrew Jackson's army, studied law, was elected to Congress, and became

Sam Houston beat Santa Anna with an army half the size of the Mexican general's.

San Antonio was the center of Spanish society in Texas even when Mexico no longer belonged to Spain. Mexicans still danced the fandango in their Spanish-style clothes at balls in 1844, when a European artist painted this scene.

Most Texans wanted to be part of the United States. But neither President Jackson nor President Van Buren wanted to let Texas in. They were afraid of more trouble with Mexico if they did. And Texas wanted to be a slave state—that would upset the balance in Congress.

governor of Tennessee. But when his young wife left him he was in despair. He became an Indian again—this time for four years—and then moved to Texas as an Indian trader.

Sam Houston and his followers decided to fight Santa Anna at San Jacinto, which is near the city that is now called Houston. It was April of 1836; the Texans were outnumbered, but they were smart. They waited until the siesta hour, which is a time after lunch when some Spaniards and Mexicans take a nap. It didn't take long—some say only 15 minutes—and Houston and his men captured Santa Anna and routed the Mexicans. They made Santa Anna sign a treaty that made Texas an independent nation. Sam Houston was elected president of the new nation: the republic of Texas. Texas had its own flag, with one lone star on it.

Susanna Dickinson was a young wife with a baby when her husband died defending the Alamo. She was spared so she could warn Houston about the fate of those who defied Santa Anna.

Houston wanted Texas to become part of the United States. There should have been no problem with that, except that some Texans wanted to have slaves. By this time, the United States was divided: there were slave states and free states, and they were equal in number. If Texas became a state, and a slave state, the South would have more votes in Congress than the North. That would create trouble. President Andrew Jackson had to say no to his old soldier friend Sam Houston. Texas stayed independent.

Finally, in December of 1845, while James K. Polk was president, Congress made Texas a state (the 28th state). Sam Houston became a senator from Texas.

By this time, slavery was becoming a

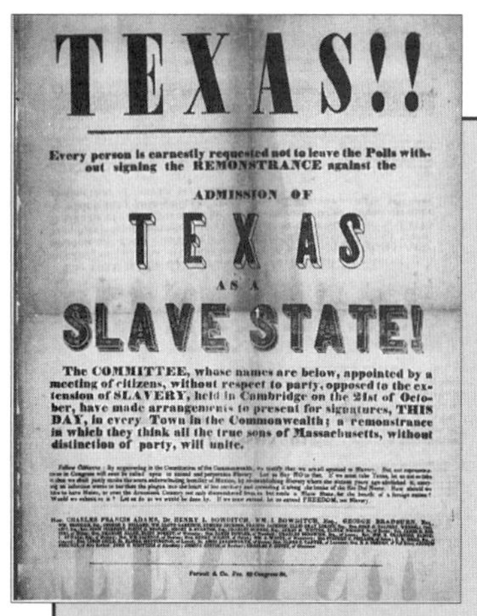

This poster urged Massachusetts voters to protest against admitting Texas into the Union as a slave state.

A Compromise, Not a Solution

The problem of keeping a balance between slave and free states had come up when Missouri entered the union as a slave state in 1820. Maine became a free state at the same time, so that there would be an equal number of slave and free states. That was known as the *Missouri Compromise*, and it was a law passed by Congress in 1820. It did more than just create two new states. The Missouri Compromise said that the rest of the territory left from the Louisiana Purchase would be free territory (and, after that, free states). For 30 years that idea seemed reasonable; then something happened. The Compromise was broken. Read on, and you'll see how that led to trouble—big trouble.

hot issue. The abolitionists were trying to end slavery; the slave owners were trying to convince everyone that slavery was a good thing—they wanted the United States to become a slave nation. When they realized that wasn't going to happen, many Texans and other Southerners began saying that the South should go its own way, and that the Southern states should secede (suh-SEED) from the United States and become a separate nation—a nation built on slavery.

Sam Houston disagreed. He had worked hard to make Texas a part of the Union. He hated the idea of secession and he didn't like slavery. It took courage to say what he thought. Especially in 1859, which was when he was elected governor again. His ideas got him kicked out of the governor's office. But that happened in 1861. Right now it is 1846 and trouble is brewing on the Mexican border.

This is an artist's idea of Davy Crockett's final moments at the Alamo. It's exciting but probably didn't happen this way—the defenders' last stand was inside the mission buildings.

10 Fighting Over a Border

Zachary Taylor entered the U.S. army in 1808, when he was 18, and served for 39 years.

The Texans thought their southern border went down to the Rio Grande river. Mexico said, "No, it doesn't." The U.S. said that the Mexicans owed a lot of money to American citizens and it was time to pay up. Things got tense. Both countries sent armies to the Texas border. Both those armies had hotheads. President James K. Polk had already decided to declare war, when some Mexicans killed some Americans (or so it was said). It was May of 1846, and the Mexican War had begun.

People in the United States had mixed feelings about the war. President Polk was eager to fight. So were many other people—what they wanted was Mexican land. Thousands rushed to volunteer. Some people thought the United States should take all of Mexico. Slave owners saw Mexico as a place to extend slavery. Some said they wanted to spread the American way of life. It was that manifest destiny idea.

But not everyone agreed. Some Americans said the United States was acting like a bully, picking on a weak neighbor. It took courage to speak out against the war; it usually does. Frederick Douglass, the abolitionist, had courage. Douglass wrote:

> *In our judgment, those who have all along been loudly in favor of...the war, and heralding its bloody triumphs with apparent rapture...have succeeded in robbing Mexico of her territory....we are not the people to rejoice; we ought rather blush and hang our heads for shame.*

Writer Henry David Thoreau refused to pay his taxes and went to jail. (He didn't want his tax money used to support a war.) Three church groups—Congregational, Quaker, and Unitarian—denounced

General Zachary Taylor wins one battle after another and is soon a national hero. Taylor is a Whig. So is that other soldier-hero, General Winfield Scott. President Polk, a Democrat, is not pleased with their acclaim. He knows that the last soldier-hero, Andrew Jackson, was swept into the presidency. (Polk is right to worry; Taylor will become president in 1849.)

Back East, most Americans supported the war and thirsted for bulletins; Richard Caton Woodville called his picture *War News from Mexico.*

the war. Walt Whitman, a journalist who would soon become a poet, wrote that "America knows how to crush, as well as how to expand." (He favored peaceful expansion.) From the war front there were soon reports of ransacked towns, drunken soldiers, and senseless killing of civilians. (Wars are usually like that.) More than 9,000 soldiers deserted the army before the war was over.

Henry Clay, who had been a war hawk in 1812, wrote, *This is no war of defense, but one of unnecessary and offensive aggression. It is*

One of the United States divisions marches through San Antonio's Grand Military Plaza at the start of a 900-mile trek into Mexico.

Mexico that is defending her firesides...not we. A gangly, long-legged, 38-year-old congressman named Abraham Lincoln stood up in Congress and attacked President Polk for starting an unnecessary war. *Allow the president to invade a neighboring nation...whenever he may choose to...and you allow him to make war at pleasure.* The House of Representatives passed a resolution condemning Polk.

But in New York, Philadelphia, Indianapolis, and many other places, the war was very popular. It was a war fought for territory, not ideals; most Americans, at the time, seemed to want it that way. At parades and rallies, citizens cheered the war effort. American soldiers fought all the way to Mexico City, to "the halls of Montezuma." They fought Santa Anna—who was back in power—and they won the war.

Many of the soldiers

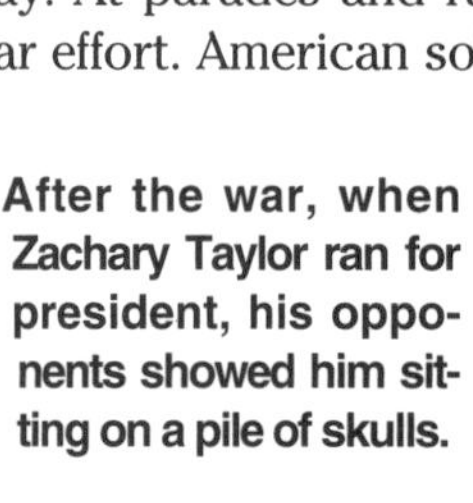

After the war, when Zachary Taylor ran for president, his opponents showed him sitting on a pile of skulls.

who fought together in Mexico would soon be fighting against each other. (Some of their names are: Robert E. Lee, T. J. "Stonewall" Jackson, Ulysses S. Grant, and George B. McClellan.) None of them realized it then, but the Mexican War was a training school for another war that was soon to come.

General Winfield Scott, who was called "Old Fuss and Feathers" because he always looked splendid in his neat, elegant uniform, became a national hero. So did another general, "Old Rough and Ready" Zachary Taylor, whose uniform wasn't neat at all.

Soldiers and sailors returned home singing this song:

When Zacharias Taylor gained the day,
Heave away, Santy Anno;
He made poor Santy run away,
All on the plains of Mexico.

(Chorus)
So heave her up and away we'll go,
Heave away, Santy Anno;
Heave her up and away we'll go,
All on the plains of Mexico.

General Scott and Taylor, too,
Heave away, Santy Anno;
Have made poor Santy meet his Waterloo,
All on the plains of Mexico.

(Repeat Chorus)

When the war was over, the Texas–Mexico border was set at the

The Mexican military academy was housed in a fort named Chapultepec. General Winfield Scott (above) attacked, and the Mexican army and the boy cadets fought him hand-to-hand. When Americans charged Molino del Rey (below—it was part of the fortress's defenses), half of them were killed. But Scott took Chapultepec and became a hero in the States.

A young lieutenant—his name was Ulysses S. Grant—said the Mexican War was "one of the most unjust ever waged by a stronger against a weaker nation."

Rio Grande river. But the United States got more than that border settlement; under the terms of the treaty of Guadalupe Hidalgo, signed in 1848, the United States received California—which at the time meant land that stretched from Texas to California and went as far north as Wyoming.

President Polk didn't know how lucky he was. Nor did the rest of the nation. Something had just been discovered in California on land that belonged to that good-hearted Swiss-Californian, John A. Sutter. Something very valuable.

Like most of the inhabitants of the old Spanish lands, the Californians were great horsemen who fought with lassos and lances.

The Californians had no close ties to Mexico, but when Americans in California marched all over proclaiming a republic, the Californians got mad and fought back. The last fight was at the Plains of Mesa, near Los Angeles, in 1847. It wasn't as bloody as it looks here.

11 There's Gold in Them Hills

After his discovery, Marshall couldn't prospect for gold in peace. Other miners thought he was lucky and hung around him all the time.

It wasn't Cíbola, it was California—but it had gold! The gold the Spaniards, the French, and the English had all sought so desperately.

It had been there all along—shiny and pretty—right in California's mountain streams, where a carpenter named James Marshall found it without even looking for it. He was building a mill for John Sutter when, on January 24, 1848, he came across some heavy golden flakes. Right away, he thought he knew what they were, but he took them to his boss to make sure. Sutter tested them, then told Marshall to try and keep them a secret. It was too late for that. The workers at the mill already knew, and Sutter and Marshall did some bragging, too.

"Any man who makes a trip by land to California deserves to find a fortune," wrote one man in a letter home.

Nine days after Marshall's discovery the United States signed the Treaty of Guadalupe Hidalgo. That ended the war with Mexico and gave California to the United States. It took a little longer than that for people in the States and Mexico and the rest of the world to know about California's gold, and a bit of time for them to believe it wasn't a hoax—that it was really true. But that didn't take too long, either.

This is Sutter's Mill in 1851; the man in front is James Marshall. In 1852 a hopeful miner arrived and wrote: "To our left is Sutter's Fort, an ancient and dilapidated-looking concern, all gone, or going to decay."

By "extreme clipper ship" to San Francisco: "This splendid vessel is one of the fastest clippers afloat, and a great favorite with all shippers."

Once they started believing, they just tossed aside whatever they were doing and headed for California.

Farmers left their plows, blacksmiths left their forges, tailors left their needles, sailors left their ships, and doctors left their patients. Where did they all go? Why, to the goldfields, of course.

Some of them prepared for the trip about the way the first gentleman voyagers prepared for their trip to the New World. They got out their best clothes and a few fancy tools and headed west. Those men—and most of those who went to the goldfields were men—usually didn't make it. Some died of disease, or fatigue, or Indian attack, or rattlesnake bite, or starvation. Some froze to death in the Sierra Nevada mountains, or died of thirst in the desert. But those who took their time and planned the trip—and were lucky—made it.

There were three ways to get to California from the East. One way was by ship down around Cape Horn at the tip of South America. It was a long trip, and if the ship didn't carry proper provisions, you might have scurvy (the miserable disease that comes from a lack of vitamin C) when you arrived. Still, it may have been the easiest way.

Or you could go by boat to Panama, then overland and upriver to the Pacific Coast, where you waited for another boat to take you north. That was the fastest way to go—if you didn't get robbed or killed in Panama, or catch malaria (another killer).

The cheapest way to go, and so the route most people took, was overland. That was difficult—you know that—but it got a little easier each year. At least, it was easier if you had enough sense to get to California before the snows made the Sierra Nevada just about impassable. (I'm not going to tell you what happened to Virginia Reed's parents and the Donner party when they tried to get over the mountains in

Pancakes (the miners called them flapjacks) were everyday food in the mining camps. "We became expert in flapping them over in a frying pan," said miner Lemuel McKeeby. "It wasn't difficult for me to throw them some two or three feet up in the air and land them safely, batterside down, in the pan. I heard some experts in this line who claimed that they could throw them up the chimney, then run around on the outside of the cabin and catch them in the pan!"

"The dead teams of '49 and '50 were seen scattered...upon the road. Very soon...they...filled the entire roadside; mostly oxen, here and there a horse and once in a while a mule. Wagons, wagon irons, ox chains, harness, rifles..."

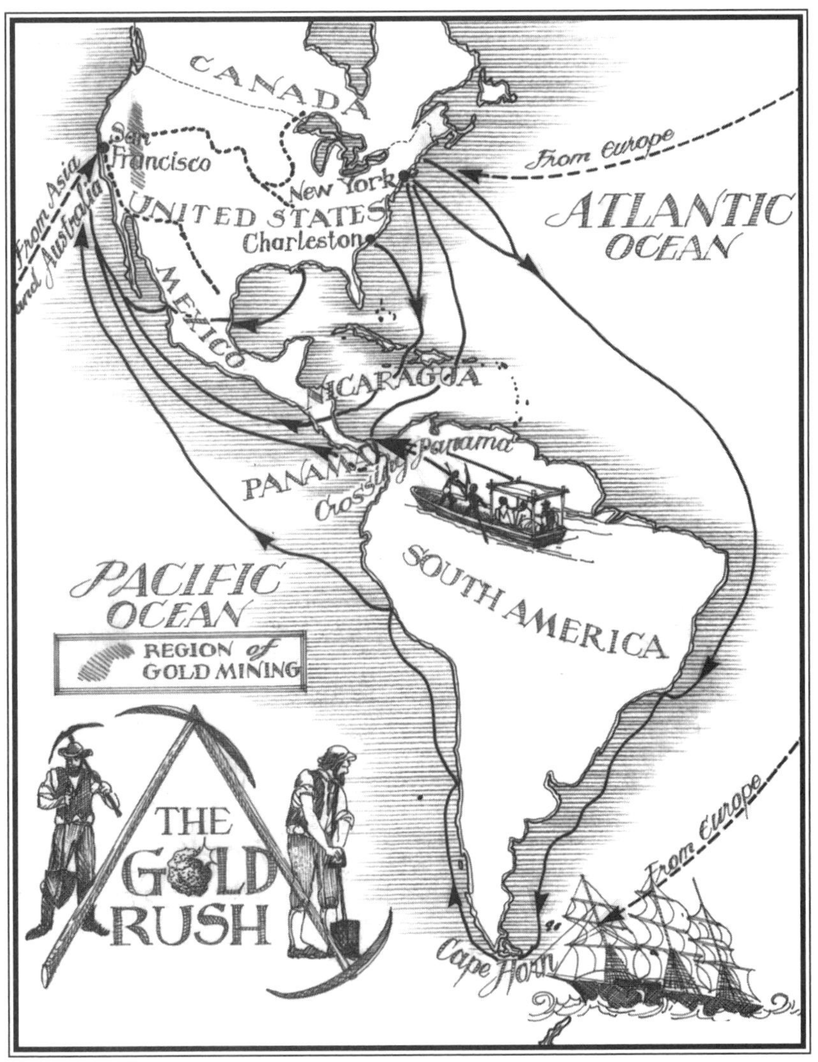

winter. Look up *Donner* in an encyclopedia, if you like grisly stories.)

Once you were in California, your cares would be over. Why, you could bend down and pick up gold in the streams. You just needed a strong sack and a strong back to haul it all down from the hills. Sure. Well, that's what they said back East. Nothing to it, they said. Sell everything you have and head west. You'll be rich in no time at all. Sure.

Guess who got rich? The storekeepers. The people who sold things to all those miners who poured into California. The prospectors (that's what the miners were called) all had to eat. And they needed tents and shovels and shirts and pants. There wasn't much available in California in the way of supplies or food.

In 1859, miner John H. Gregory hits pay dirt (gold) in the western part of Kansas territory (today it's Colorado). Newspaper man Horace Greeley is traveling in the area and sends word back east. That does it. By 1860, some 5,000 miners a *week* are pouring into the area. Reaching the ore in the Rockies takes heavy drilling equipment. Most of the miners go home broke. But towns like Boulder, Golden, Carson City, and Colorado Springs are born during the mining fury.

Forty-niners young and old. Being photographed with the miner's tools of trade became a popular fad—some had themselves snapped without ever setting foot in a mine.

Oh, Sally, dearest Sally!
Oh, Sally, fer your sake,
I'll go to Californy,
And try to raise a stake.

—"JOE BOWERS FROM PIKE," A POPULAR GOLD RUSH SONG

So things got mighty expensive. There's an economic law called the *law of supply and demand.* It's a common sense law. Here it is: the less there is of something that everyone wants, the more it will cost (and vice versa).

In 1848 there were 812 people living in San Francisco. Two years later, San Francisco was a city of 25,000. The mining areas went from no one to thousands. In 1848, 400 settlers arrived in California looking for land. In 1849, when the news of gold was out, 80,000 are said to have arrived looking for gold.

No one was prepared for all those people. So an egg that might have sold for a penny or two in the East sometimes sold for as much as a dollar in California. You can guess what happened to room rents in San Francisco.

Oh, there were a few miners who did strike it rich. A few did reach down with a knife or spoon and find nuggets of gold in the mountain streams; some became wealthy. But not many. Louisa Clappe, who was one of the few women in the mining camps, wrote:

> *Gold mining is nature's great lottery scheme. A man may work in a claim for many months, and be poorer at the end of the time than when he commenced; or he may take out thousands in a few hours. It is a mere matter of chance.*

The Chinese, the Know-

It was 1849, and the sailors on the Yankee vessels stopping at Canton and Hong Kong had astounding news. They told of gold, mountains of gold, streams filled with gold, all in a place called California. The gold was there for anyone to take, the sailors said. The Chinese needed to hear good news. Wars and rebellions had disrupted the ancient Chinese ways. Many people were out of work.

So they took their picks and shovels and headed for California. They didn't intend to stay. They would just reach down and get some of that gold and come home rich. (Many other immigrant groups—the Italians and the Irish, for instance—had the same idea. They would come to America, get rich, and go back to the old country. Some did just that; but most stayed.)

In 1852, more than 20,000 Chinese passed through the San Francisco Customs House on their way to the goldfields. And that was just the beginning. Soon there was a big Chinese population in San Francisco. Chinese miners were working claims in the Sierra Nevadas; Chinese merchants were importing and selling Chinese food and goods; and Chinese opera companies were entertaining theatergoers. (In 1853, the promoter P. T. Barnum brought the Hong Fook Tong opera company to New York, where the *New York Herald* wrote of "magnificent costumes, antique style...and extraordinary movements on the stage.")

Nothings, and Abraham Lincoln

Some of the Chinese miners did find gold. There was plenty of work for the others. Since there were few women settlers in California, there was much cooking and laundry to be done. Most American men wouldn't do those traditionally female jobs; Chinese custom held it no disgrace to cook and clean.

At first, the Chinese were treated like the other new Californians. It was an exciting time, and gold was beckoning. But when disappointment set in—most people didn't find gold—someone needed to be blamed. So why not blame men who spoke a foreign tongue and wore their hair braided in long pigtails?

Soon there were nasty attacks on the Chinese, and discriminatory laws that were another form of attack. Partly this was because of an idea, called *nativism*, that was leaving some Americans confused. The nativists said that only white Anglo-Saxon Protestants were "real" Americans. The nativists wanted to keep most other people from entering the country. And they wanted to keep blacks enslaved. What was really strange was that the nativists didn't even think that Native Americans belonged in America! On the West Coast the nativists attacked the Chinese. On the East Coast they attacked the newest European immigrants—especially those from Catholic countries.

The nativists even had their own political party. They called it the American Party. But others called it the "Know-Nothing Party." Don't laugh. The Know-Nothings actually elected six governors.

These men and boys of the West found watching Chinese eat with chopsticks as fascinating as we would find going to the movies. They didn't ask the Chinese if they minded being watched.

In 1855, an Illinois lawyer named Abraham Lincoln wrote a letter to a friend saying just what he thought of the Know-Nothing Party. This is what Lincoln wrote:

> *As a nation we began by declaring that "all men are created equal." We now practically read it, "all men are created equal, except Negroes." When the Know-Nothings get control, it will read "all men are created equal except Negroes and foreigners and Catholics." When it comes to this, I shall prefer emigrating to some country where they make no pretense of loving liberty.*

But Abraham Lincoln didn't have to emigrate; most Americans rejected the nativist ideas.

Before he brought over a whole opera company, P. T. Barnum made money with a simpler idea: he sent an ordinary Chinese family touring the country with his traveling show.

I have seen purer liquors, better segars, finer tobacco, truer guns and pistols, larger dirks and bowie knives...here in San Francisco, than in any other place I have visited; and it is my unbiased opinion that California can and does furnish the best bad things that are obtainable in America.

—HINTON R. HELPER, *LAND OF GOLD: REALITY VERSUS FICTION*, 1855

Many of the Forty-niners, as those who came that first year were called, did find gold, but usually it was gold dust and gold flakes. It often took all they found just to eat.

At first they panned for gold. They washed gravel and sand from the streams. If there was gold it sank to the bottom of the pan, because gold is heavier than sand. Soon most of that surface gold was gone. Mining became much harder.

But it was the stories of the lucky miners that filled the newspapers and that kept people coming. They didn't just come from the States—although most did. They came from Mexico, Spain, China, Peru, England; you name a place, and someone was bound to be from there. They all had gold fever—it was contagious.

In Norway a newspaper printed a letter from California; here is part of it:

> *A pair of boots that cost $2.50 in New York are $20 here. A pair of shoes that were 75 cents in New York are $8 here, and so on. These are high prices, to be sure, but if you work hard you can still make money. In May, I saved $223; in June, $295. Yesterday alone I made $35. All my earnings from May 1 to July 14 amount to $750, which is $120 more than the cost of the journey here.*
>
> *The work is extremely hard. I start at 4 o'clock in the morning and keep on till 12 noon. After that I rest for three or four hours, for at that time of day the heat is unbearable, and then I work again till 8 o'clock in the evening. The nights here are exceedingly cold. We live in tents; I have not been inside a house since April 1. The ground is our bed and a saddle or something like that our pillow.*
>
> *This kind of life agrees with me and my health is excellent. We live a free life, and the best thing of all, that which I have always considered one of the supreme blessings of existence, is that no human being here sets himself up as your lord and master....*
>
> *The gold we find is almost completely pure. The size of the nuggets varies. In some places pieces have been found that weighed up to seven pounds. Here, at the river where I am staying, it is found almost like fish scales, very thin....You obtain it by washing out the dirt in a machine which looks like a roller and that is what it is called. You throw the dirt in one end of the machine, which is somewhat higher than the other, and start the machine, all the time adding a certain*

The Rare Sex

Lots of things were in short supply in the mining camps, but there was something the men missed most of all—women. So when one appeared, men would come from miles around just to hear a female voice. When Louisa Clappe arrived with her husband (he was a doctor) in the Rich Bar camp, one prospector (he was from Georgia) rushed into the doctor's office with a bottle of champagne. He said he hadn't seen a woman in two years. Said Louisa, "I assisted the company in drinking to the honor of my own arrival."

"There are but few women...we have not seen more than 10 or 12."

quantity of water. By this process, lighter particles, like dirt and pebbles, are washed away, and the gold is left behind.

That letter was enough to bring Norwegians to California.

San Francisco had an assorted population. Sometimes all those different people got along very well. Sometimes they didn't. Sometimes things got rough. No one was in charge in California. The Mexican authorities were out of power and the Americans were just getting organized. (Although, in 1849, Californians did call a convention, write a constitution, elect a governor, and prohibit slavery.)

In one year, San Francisco (above) changed from a tiny village to a boom town with a population of 25,000. In the early years of the gold rush, Mexican, Chinese, Yankees, and Southerners mixed and drank together in the saloons.

What about James Marshall and John Sutter? Did they get rich? No. Gold made them both poor. John Sutter was afraid of gold. He knew what gold greed could do to people. It was worse than he imagined. People trampled his property, ate his cattle, and left him ruined. James Marshall never profited from the gold either; he began to drink and died a sad man.

California wasn't the only place where

Prospectors washing gold from pay dirt in a "long tom"on the American River. Said an observer, "The machinery is very simple, being an ordinary trough made of plank...with a sieve at one end to catch the larger gravel, and three or four small bars across the bottom...to keep the gold from going out with the dirt and water at the lower end."

In 1861, Orion Clemens is named secretary of the Territory of Nevada. He takes his younger brother, Samuel Clemens, with him as an assistant. They head for Carson City; there Sam (left, aged 15) changes his name to Mark Twain and becomes "smitten with silver fever." Later, Twain writes of his mining experiences in a book called *Roughing It.*

gold was found. It also turned up in Oregon, Nevada, Wyoming, Montana, and Colorado. Silver, too, in some of those places. Now why hadn't the Spaniards, with all their looking, ever found it? Well, it seems that a few of them may have done so. (This is part of what makes history so interesting. We keep discovering new things about the past.) Archaeologists—those professional diggers—have found the ruins of old mines in the Rocky Mountains; they think they were Spanish gold mines. Why didn't anyone know about them? Probably because the Spanish government, back in the 16th and 17th centuries, took one fifth of the profits from any gold mines for the royal treasury. Now, if you found a gold mine, and you were a Spanish subject, would you tell anyone about it?

Back to California. What did gold do to California? It brought great wealth to a quiet frontier. It brought people and ideas from around the world. It mixed rich and poor. It gave California enough people to become a state—quickly—and that is only part of what it did.

They swam the wide rivers and crossed the tall peaks
And camped on the prairie for weeks upon weeks.
Starvation and cholera and hard work and slaughter,
They reached California spite hell and high water.

James Marshall died a poor man, but after he was dead, things changed for him. The Society of California Pioneers and the Native Sons of the Golden West fought over his body—both of them wanted to bury it. They had to keep it packed in ice until they agreed to do it together. Then they buried him in style, spent thousands of dollars on a statue of him, and set up a fund to keep it clean for the tourists to admire.

What Was Your Name in the States?

The richest mining discovery in U.S. history was the legendary Comstock Lode at Virginia City, Nevada. It was actually discovered by two Irish prospectors, Peter O'Reiley and Pat McLaughlin, but Henry T. P. Comstock came along and kind of bamboozled them into a partnership arrangement. "Old Pancake" (that was Comstock's nickname) became famous and rich, for a while. Old Pancake was a good-natured guy (when he wasn't drunk), and foolish (when he was). He managed to go through his riches and finally shot himself in the head.

A Comstock miner (this was one of the first American flash photographs).

The Comstock Lode had fabulous gold and even more fabulous silver veins. Prospectors poured into the area. They had to live somewhere, so Virginia City sprang up to house the miners. It was boisterous and busy. John Ross Browne, who was there, described it: *Steam-engines are puffing off their steam; smoke-stacks are blackening the air with their thick volumes of smoke; quartz-batteries are battering; hammers are hammering; subterranean blasts are bursting up the earth; picks and crowbars are picking and crashing into the precious rocks; shanties are springing up, and carpenters are sawing and ripping and nailing.* And then there were the people; prospectors, shopkeepers, peddlers, saloon-keepers, barroom girls, cooks, wagoneers, and others who were just there for the excitement.

Virginia City? Where did it get that name? Well, the story is told of a miner named James "Old Virginny" Finney, who fell down drunk one night and broke his bottle. When he got up he said he'd baptized the ground "Virginia," after himself.

Bringing ore up at the Comstock. Tools and equipment were basic at best.

12 Clipper Ships and Pony Express

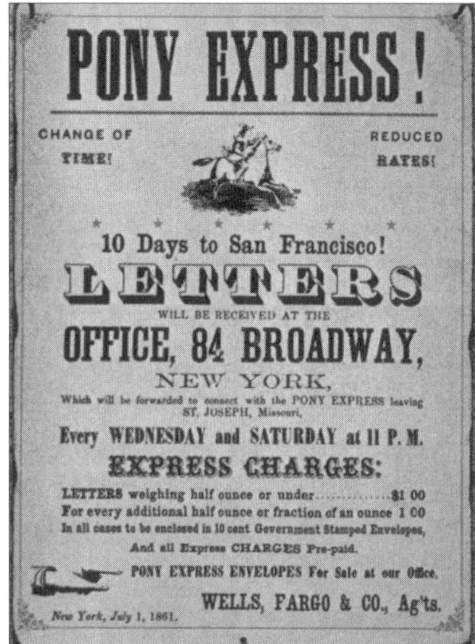

The Pony Express mail was wrapped in oiled silk and carried in four leather bags attached to the saddle; 20 pounds was the maximum weight one rider carried.

In 1850 the 31st star was sewn on the American flag. The star was put there for the new state of California. Do you remember how many states there were in 1789? How many are there now? When did your state become a part of the Union? Do you know when Utah became a state? Or Arizona? Most places in the West were territories for a long time before they became states. California leapfrogged into statehood. That was because so many people moved to California so fast. That created problems, but also opportunities.

Levi Strauss saw an opportunity. He realized that miners needed sturdy pants. He bought yards and yards of heavy canvas and turned it into strong pants, with rivets to hold the seams in place. Those pants were called *Levi's*, and still are.

Many of the miners were homesick. They wanted to know what was happening to their loved ones back home in Ohio, or New Hampshire, or Alabama. And the people back home were frantic. Had their fathers and brothers made it to the

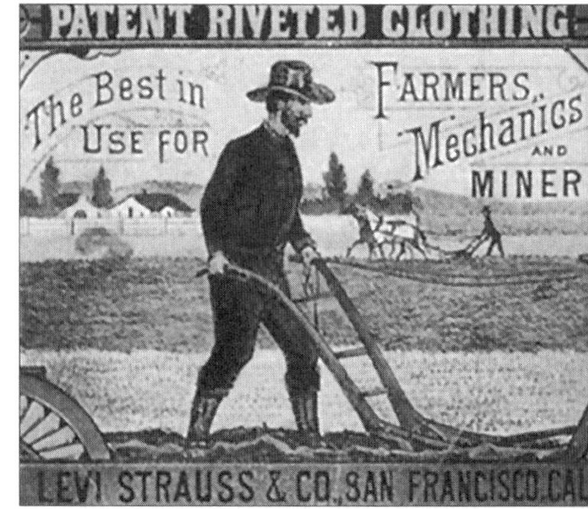

Levi Strauss meant to sell canvas for tents; but he did better when he made the canvas into pants. Miners liked lots of pockets to keep their tools handy.

goldfields? Sometimes a year went by without letters or news. Was Father dead of cholera? Was he rich? Or was he too embarrassed to come home poorer than when he left?

The United States Post Office in California was not prepared for the thousands and thousands of letters that came in 1849. There weren't enough postal workers, and those doing the job kept quitting to work in the goldfields. No one was around to deliver the mail.

Back East, many people were anxious to get to California. But they wanted a safe, quick means of travel. Americans used their ingenuity to solve those problems.

Soon there were private postal services. (One was called the Pony Express.) Then came telegraph lines; they carried messages with the wizardry of electricity.

Stagecoaches, changing horses at outposts about every 20 miles, soon whizzed people across the country. (Well, it seemed like whizzing then.)

And clipper ships really did whiz across the seas.

Sailors will tell you that the clippers were the loveliest ships ever built. Slim, graceful sailing ships with three masts and square sails, they were the largest sailboats that anyone had ever seen. Some clipper

This was the ad for Pony Express riders: "WANTED: Young skinny wirey fellows, not over 18. Must be expert riders willing to risk death daily. Orphans preferred. Wages $25 per week. Apply, Central Overland Express, Alta Building, Montgomery Street, San Francisco."

Stagecoaches got their name because they traveled in stages, changing horses at each stage.

There was a time before our time,
It will not come again,
When the best ships still were wooden ships
But the men were iron men.

From Stonington to Kennebunk
The Yankee hammers plied
To build clippers of the wave
That were New England's pride.

The "Flying Cloud," the "Northern Light,"
The "Sovereign of the Seas"
—There was salt music in the blood
That thought of names like these.

—STEPHEN VINCENT BENET

masts were tall as a 15-story building. The classy, designed-in-America clipper ships were the envy of the world. Ships had been taking eight or nine months to make the trip from Boston around Cape Horn to San Francisco. The *Flying Cloud*, one of the largest of the clipper ships, made it around the Horn in 89 days (how many months was that?). Until 1989, no sailing ship ever went faster.

However, if you were really in a hurry to get your mail across the country you could send it by an even speedier route. The fastest mail went by Pony Express. Galloping horses raced from St. Joseph, Missouri, to Sacramento, California, in an incredible 10 days. This is how the Pony Express worked: stations were set up 10 to 15 miles apart, all along the route. Horses were ready at each station. A rider starting in Sacramento rode as fast as he could to the first station. Then he jumped off his tired horse and onto a fresh one, grabbed his mail bags, and headed on. Usually it took two minutes to change horses. About every eight stations, a new rider was ready to take over. Pony Express riders rode through the night, through rain, through blizzards. They had to protect themselves from Indian attack.

A Pony Express rider.

The *Flying Cloud* was 225 feet long and 1,783 tons. On its record-breaking 5,912-mile journey around Cape Horn skippered by Captain Josiah P. Creesy, the ship averaged 227 miles a day.

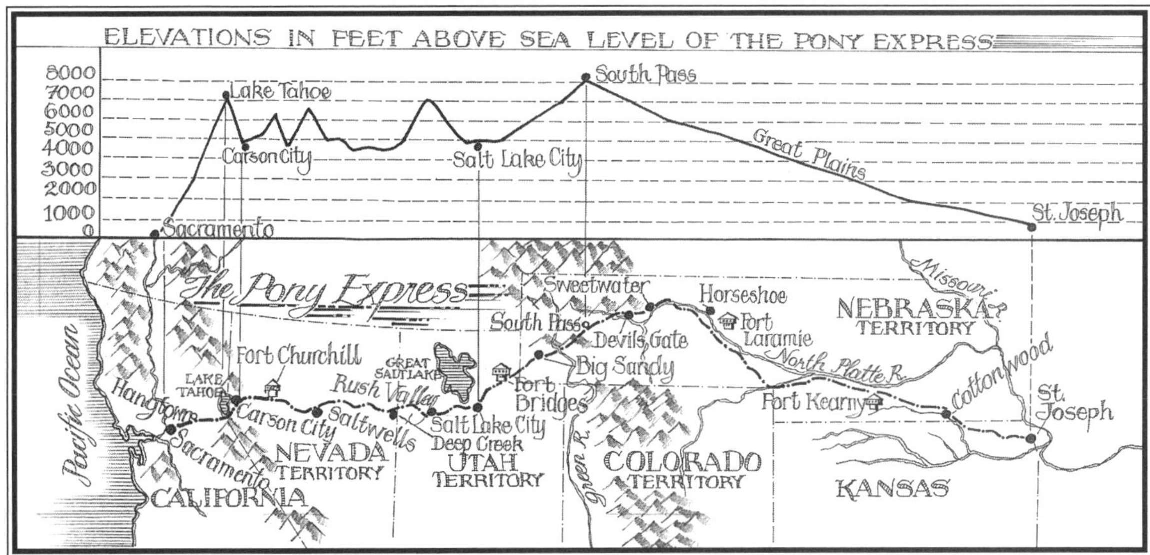

Some of them, like William "Buffalo Bill" Cody, became famous for their exploits. There were 190 Pony Express stations, 500 horses, and 80 riders.

The Pony Express riders didn't just go across the country—they went up and down it, too. The top half of the map shows the height above sea level (in feet) of stages along the route. At South Pass, where the riders crossed the Rockies, they had climbed to 8,000 feet. But the Pony Express lasted only 18 months—just until the telegraph lines (below) went up.

People who were in a hurry for news, like newspaper editors, sometimes used pigeons. Yes, real birds. They were trained to fly from place to place with small bits of paper taped to their legs. You can understand that a pigeon can't carry a whole lot of news. Samuel F. B. Morse, who was an artist and needed to earn some money, came up with a much better idea. He developed the telegraph. Using electricity, he sent messages on a wire.

You may wonder about an artist being an inventor. But in the 19th century most Americans were versatile. They could do a lot of things. About the middle of the 19th century, it seemed as if an idea volcano was erupting in the United States. Americans from all walks of life started coming up with ideas for practical new things. Samuel Morse's idea was one of the most important of his day, although at first most people thought the telegraph

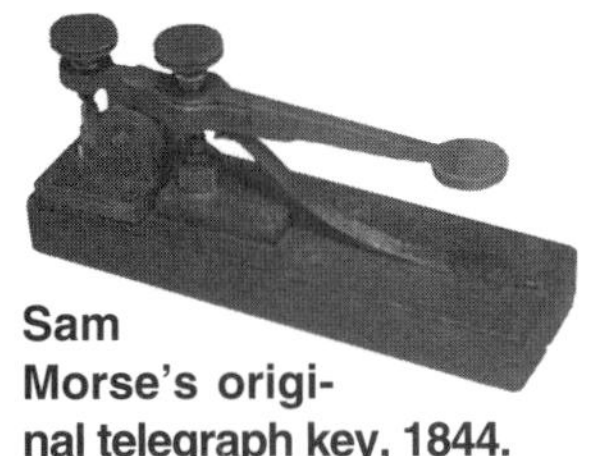

Sam Morse's original telegraph key, 1844.

In 1858 two ships meet in the middle of the ocean. One has a telegraph line brought from Newfoundland; the other a line from Ireland. The wires are spliced together, and for three weeks telegraph messages are sent from continent to continent. Then the wire fails—but the idea has been proved. Another transatlantic cable will be laid after the Civil War.

When Samuel F. B. Morse *(right)* flicked the switch and the first message was sent by telegraph, someone said that Maine could now talk to Florida. "Yes, but has Maine anything to say to Florida?" joked the writer Ralph Waldo Emerson.

was just an amusing toy.

Sam Morse thought that it might be more than a toy. He got interested in the idea on a ship coming home from Europe (where he'd been studying painting). Someone told him about Joseph Henry's work. (Henry was the greatest American scientist after Benjamin Franklin. He did important work in electromagnetism and experimented with a crude telegraph.) Morse wondered if electrical pulses could be sent along a wire in dots and dashes. If so, those dots and dashes could be used as a code to carry messages. Starting with that simple idea, Morse devised the telegraph. It took 12 years to work out the details and get wires strung on poles, but it happened. In 1844 the first message went—by Morse code—from the Supreme Court in Washington to Baltimore. "What hath God wrought?" asked Morse by telegraph. A few seconds later, a message was sent the other way.

Was there any important use for this invention? Well, seconds after Senator Henry Clay was nominated in Baltimore, Maryland, as the Whig candidate for president, people in Washington, D.C., knew about it. Astounding! Now there were those—especially newspaper editors—who began to see that the telegraph was more than a toy. Seventeen years after that first message was sent, the telegraph reached from coast to coast and the Pony Express was out of business.

When stagecoaches crossed desert, they were pulled by mules, which stood the heat better than horses. This line was named for the mules: the Jackass Mail.

13 Flying by Stagecoach

The bright-yellow Concord coach was made of ash-wood and iron, and had a low center of gravity that helped stop it tipping.

In 1860, 15-year-old Rebecca Yokum put on a new hat with a wide brim and lots of flowers, and climbed aboard a stagecoach that pulled right up to her grandfather's gate in Springfield, Missouri. Men sat back to back on the top of the stage; more men were inside; six horses were hitched to the front. One bench seat inside was reserved for Rebecca, her mother, and her three brothers.

Everything on the stage was so shiny and new that we felt we were in a palace, Rebecca recalled later. The body of the stagecoach, made with care and pride in Concord, New Hampshire, was suspended on broad

A Buffalo Charge

First came a murmuring and a new cloud on the horizon, then a swelling of sound until the earth shook with the pounding of hooves of ten thousand bison, twenty, thirty thousand furry locomotives, wild and red of eye and foaming at the mouth, terrified and contagious with terror, and they would keep coming for hours at a time, the noise and dust and madness accumulating until the human skull could scarcely keep it all out and the brain was penetrated. Then they would be gone, leaving the earth still trembling, and the people trembling, too.

—Roger Kennedy, *Rediscovering America*

Fitz Hugh Ludlow rode the Overland Mail and said, "It was...like being in an armchair, and sentenced not to get out of it from Missouri to California....The back cushions of the wagon were stuffed as hard as cricket-balls, and the seat might have been the flat side of a bat. I tried fastening my head in a corner by a pocket-handkerchief sling; but just as unconsciousness arrived, the head was sure to slip out, and, in despair, I finally gave over trying to do anything with it."

leather straps that absorbed some of the jolting taken by the wide, iron-bound wooden wheels. It was painted canary yellow.

Rebecca's father, a Presbyterian minister, had gone to California in 1859. He wanted his family to join him; he told them to come by train and coach across the Isthmus of Panama and then by boat to California. *Don't come by stage under any circumstances,* Rebecca's father wrote.

But Rebecca's mother had a mind of her own.

> *Mother...was a very determined woman and...she prepared to set out without delay for California on the Butterfield Overland stages.*

The Butterfield Overland Mail Company was born in September 1858. There was a need for regular mail service for letters and packages. The Pony Express was expensive; it wasn't for ordinary letters. The telegraph, when it was completed, handled only important messages that needed quick delivery.

Trains from the East now chugged as far as Tipton, Missouri, 160 miles west of St. Louis. Mail that went by train from New York to Missouri, and then on by stagecoach, reached California in about four weeks. Not like a

Keeping Up With the Presidents and Other Things

A Presidential Review

☞ **While Jed Smith** was trekking to California, and Susan Magoffin was walking the Santa Fe Trail, in Washington, D.C., the presidents and congress were governing. Have you forgotten the presidents? We had eight of them during the first 51 years of our history. And eight more during the next 20 years. Do you remember the first eight? They are the easy ones. See if you can say their names. *Don't look at the next paragraph until you try.*

You don't expect me to tell you the name of the first president, do you? Of course you know the name of the father of our country. And you probably know that *John Adams* came next, followed by *Thomas Jefferson, James Madison*, and *James Monroe*. But did you remember *John Quincy Adams, Andrew Jackson*, and *Martin Van Buren?*

J. Q. Adams

Old Hickory

☞ **In 1824,** *John Quincy Adams* barely beat *Andrew Jackson* for the presidency. It was while *JQA* was president that Jedediah Smith went overland from the Great Salt Lake in Utah all the way to San Diego, California. John James Audubon published the first volume of *Birds of America* about the same time. (You'll read about Audubon in this book.)

☞ **In 1828**, *Jackson* beat *Adams*, and now he was president. *Jackson*, who was known as Old Hickory, was one of our most popular presidents. But not everyone liked him. A new political party, the Whig Party, was founded by the anti-Jacksonians.

☞ **In 1837** New Yorker *Martin Van Buren*—who was *Jackson's* vice president—became president number eight. *President Jackson* and *President-to-be Van Buren* rode to the Capitol together for *Van* telegraph, but not bad at all!

It was the mail that made the stagecoaches profitable; but people benefited too. The stagecoaches carried passengers as well as mail. Now that California was a state, a lot of people wanted to get there. Gold wasn't the only attraction. Some wanted to teach, some to preach, some wanted to farm, some wanted to go into business.

The stage ran night and day, with 10-minute stops to change horses. Rebecca remembered:

> *The moment we rolled into the station the tired horses were dragged away and fresh ones put in their places. Speed seemed to be the one and only thing the stage people desired.*

At one stage station they were told that a day or two before,

An Indian attack. One man, newly arrived in what was to be Wyoming, wrote: "At the time of my arrival...the Indians would fight to the death for home and native land."

Van Buren

Buren's inauguration. *Jackson* sat on the right of the carriage. After the ceremony they reversed seats for the return trip. That has been the custom ever since.

☞ **The United States** had three presidents in 1841. *Martin Van Buren's* term ended on March 4, and that day *William Henry Harrison* became the new president. A month later, *Harrison* was dead and *John Tyler* was president.

Polk

☞ **Who were presidents** nine through 16? Can you name them? You just learned numbers nine and 10. Some historians say President 11 was a great president. (Some don't.) He was *James K. Polk*, who served from 1845 to 1849 and added the big Oregon and California territories to the United States. Almost everyone agrees that our 16th president, *Abraham Lincoln*, who took office in 1861, was the greatest. The others? You may have to work hard to remember them. Most weren't memorable.

☞ **Here is the second** group of eight: *William Henry Harrison, John Tyler, James K. Polk, Zachary Taylor, Millard Fillmore, Franklin Pierce, James Buchanan, and Abraham Lincoln. Tyler* and *Fillmore* were "accidental presidents." They were vice presidents who took over when the president died.

Tyler

Freedom of Religion

☞ **John Tyler** was the president who had the republic of Texas annexed to the Union as a slave state in 1844. But he had this to say about freedom of the mind:

"The United States have adventured upon a great and noble experiment, which is believed to have been

(continued on the next page)

Other Things *(continued)*

hazarded in the absence of all previous precedent—that of total separation of Church and State. No religious establishment *by law* exists among us. The conscience is left free from all restraint and each is permitted to worship his Maker after his own judgment. The offices of the Government are open alike to all....Such is the great experiment which we have tried, and such are the happy fruits which have resulted from it; our system of free government would be imperfect without it.

"The body may be oppressed and manacled and yet survive; but if the mind of man be fettered, its energies and faculties perish, and what remains is of the earth, earthly. Mind should be as free as the light or as the air."

Taylor

☞ **After the** Mexican War, President Zachary Taylor asked Captain William Tecumseh Sherman to explore and survey the newly acquired lands of New Mexico, Arizona, and California. Sherman spent two years crossing cactus lands, and then returned.

"Well, Captain," asked the president, "will [the new lands] pay for the blood and treasure spent in the war?"

"Well, General," said Sherman, thinking about the desert region he had explored. "Between you and me, I feel that we'll have to go to war again. Yes, we've got to have another war."

"What for?" asked the startled president.

"Why," the captain replied, "to make 'em take the darn country back!"

Indians had killed the station agents and all the passengers on a coach. It didn't happen very often. Mexicans didn't attack often either, but it did happen. There were other dangers, too.

*Somewhere at one of the stops my beautiful hat disappeared....*So did her brother's gun.

At a river, where the water was high, the Yokums got out of the stage and used driftwood to get across to the opposite bank. There, another stage was waiting.

> *Two men were helping me keep my footing on the driftwood. I got smart and jumped ahead. As I did so I slipped and fell into the water up to my waist. To add to my discomfort our trunk with all our extra clothes had been left behind several days before by the stage company.*

Rebecca never forgot the thousands and thousands of buffalo she saw, and the Indians dancing at night

The artist Alfred Jacob Miller sketched buffaloes drinking at night—and Indians cutting off the buffalo hump rib. "No traveler...to Oregon...ever forgets...the delicious flavor of the Hump rib—it is probably superior to all meats whatsoever."

around their campfires. Before long, neither the Indians nor the buffalo would be free to roam the land.

In the California mountains, the conductor of the coach played a tune on a bugle to warn anyone who might be coming the other way on the narrow mountain road. *The long notes of his bugle, echoing through the mountains, sounded very romantic.*

For the whole 21 days of the trip, Mrs. Yokum held her smallest son on her lap. The Yokums, and all the other passengers, slept sitting up, shoulder to shoulder. The men on top of the coach slept sitting there. During those three weeks, Rebecca said, they never changed their clothes. Finally, they reached California.

> *The stage people came over to the stage and inquired if there wasn't a woman and some children to get off at San Jose. Mother replied: "No. Our tickets call for Santa Clara." While the stage was still waiting, a tall, strange-looking man with whiskers all over his face, stepped out of the darkness. He poked his head through the curtains and peered into the coach. He looked us over carefully in the dim light and then said firmly: "I guess you will get off right here." Then he smiled. It was Father.*

A Whale of a Bed!

The Prince of Wales

The Prince of Wales, who was Queen Victoria's son and the future King Edward VII of England, came to Washington for a visit. Naturally, he was invited to stay at the White House. Then the question came up, "How do you treat a future king?" The White House, at that time, didn't have a decent guest bedroom. James Buchanan, who was president, agreed to give up his bedroom (and sleep on a cot outside his office). Then someone decided that the president's old bed just wouldn't do. So a huge, beautiful rosewood bed was built for the occasion. The prince, who as you can see was short, may have felt lost in that big bed. But the next inhabitant of the White House loved it. He was Abraham Lincoln, and he had never had a bed that fit his long frame so well. The bed—now called the Lincoln bed—is probably the most famous piece of furniture in White House history. Most people think it was built for Abraham Lincoln—but you know better.

14 Arithmetic at Sea

Salem's seal shows Salem's wealth. The Latin motto says: "To the farthest port of the rich East."

Nathaniel Bowditch quit school when he was 10. Now that is not something I would suggest. Nathaniel had no choice. His mother was dead and his father wasn't much good to the family: he was either drunk or away at sea.

But even though Nathaniel couldn't go to school, he didn't stop learning. He was lucky to live in Salem, Massachusetts. Do you remember Salem? It was the place where Puritans once went witch hunting and killed 20 innocent persons before they were finished. But that was old history when Bowditch was born, in 1773. Salem was a seaport, and, when Nathaniel was a boy, a boom town. Its merchants were rich, and they lived in splendid red-brick homes with white-columned entrances (it was the new Federal style of architecture). By 1830, when Andrew Jackson was president, Salem had 22,000 inhabitants and was the 10th largest city in the United States. If you figure wealth on a per-person basis, it was the richest city in the whole United States.

Ships from Salem went all over the world, but especially to China, the East Indies, and the Philippine Islands (check the map on page 92). Imagine what it was like, in those days before radios and TVs, to have a ship arrive

Salem Town

The English writer and women's rights reformer Harriet Martineau visited Salem in 1834. This is what she said about the city:

Salem, Mass....has perhaps more wealth in proportion to its population than any town in the world....These enterprising merchants...speak of Fayal and the Azores as if they were close at hand. The fruits of the Mediterranean are on every table. They have a large acquaintance at Cairo. They know Napoleon's grave at St. Helena, and have wild tales to tell of Mozambique and Madagascar....They...bring furs from the back regions of their own wide land, glance up at the Andes on their return; double Cape Horn, touch at the ports of Brazil and Guiana, look about them in the West Indies...and land some fair morning in Salem and walk home as if they had done nothing remarkable.

Bowditch (right) proved—by doing it—that ordinary seamen could learn precise navigation with compass, quadrant, and sextant (below).

in town after a three-year voyage to faraway places. Imagine the stories the sailors must have told. Imagine cargoes of spices, teas, Oriental rugs, toys, dried figs, and silks.

No wonder most Salem boys wanted to become sailors. Besides the adventure, there were riches to be had at sea.

There was also danger—many sailors never returned home. Some were swallowed by the ocean; a few were swallowed by people-eating Pacific islanders.

But the ships that returned to Salem, or Boston, or New York were filled with exotic goods that sold for big profits.

Yankee boys went to sea young—sometimes at 13 or 14. By the time they were 19 or 20 they were old sailors, and often captains of their own ships. They were trained to be merchant sailors: they learned to buy and sell as well as sail. They learned to make decisions. They had to be smart, and they were. Even the proud English admitted that the American seamen were the best in the world.

Nat Bowditch wanted to go to sea, but there didn't seem to be much chance that he would do it. Most of the Bowditch men were big and brawny, but Nathaniel was a skinny, funny-looking, lonely boy who didn't have many friends. He spent all his spare time in the library.

Salem had one of the finest libraries in the country. And, in the nearby town of Beverly, there was a famous collection of books that had been taken from a British ship captured at sea.

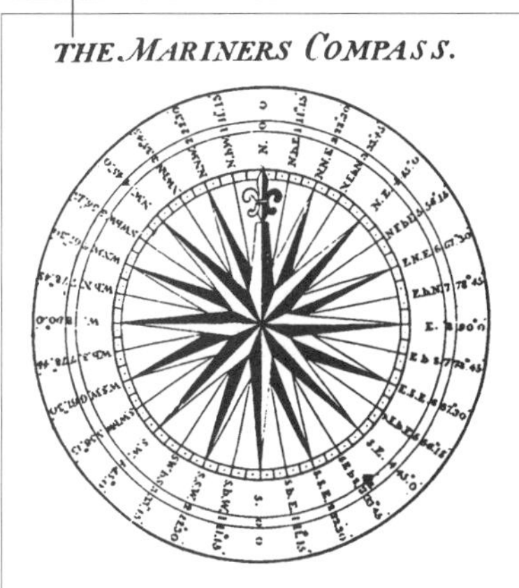

An important part of Bowditch's *New American Practical Navigator* was the mathematical tables that sailors consulted to check their instrument findings. Bowditch had combed through all the tables in Moore's *Practical Navigator*—the book that everyone had used until then—and had found over 7,000 mistakes in them. That kind of mistake could wreck a ship. Nathaniel Bowditch corrected every error.

In Bowditch's day, Salem was the hub of the China traders, and Crowninshield's Wharf (right) was the hub of Salem. The ship at the outer end of the wharf is the *America,* which in 1804 sailed into Salem harbor loaded with pepper from Sumatra. She immediately sailed out again, bound this time for Rotterdam, Holland, where the pepper brought $140,000 in gold!

Bowditch's family had him indentured to work in a chandler's, a store that sold supplies to shipowners. He had to quit school—which was hard for a boy who loved to study—but Nat's employers were kind, and he still had time to read and learn. He especially loved numbers. He taught himself everything he could about arithmetic and algebra, and then he taught himself French, so he could read the writings of the French mathematicians, who were then the best in the world. Nathaniel learned Latin and Spanish too, and read books in those languages. He had the kind of mind that wasn't happy unless it was learning.

Now sailors of his day had a big problem. When they were out at sea with no land in sight, it was hard to know exactly where they were. Sailors had been steering by the stars for centuries, but they were never able to get precise measurements. It was a problem in Columbus's time, and it had never been solved completely. How do you figure exact latitude and longitude on a rocking ship that makes instruments not quite accurate?

Bowditch figured that out. He did it with numbers and the stars. However, no one trusted his ideas, except an old school friend, Henry Prince, who was captain of a ship. Captain Prince let Nathaniel Bowditch go to sea with him and test his ideas.

In those days, Salem sailors and Boston sailors were rivals. They raced to see who could get to East Asia and back fastest. Boston

Seamen climbed the rigging with buckets of tar to tar down the ropes. It was this job that gave sailors the nickname *tars.* Sailors had to "know the ropes," too. They had to be able to put their hands on the right rope, sometimes quickly, sometimes in the dark, in order to sail the ship properly. It was better for a sailor if he wasn't very tall; quarters on board ship were cramped.

Yankee Doodles

Some say the word *Yankee* was first used by the Dutch, who called Englishmen "John Cheeses." In Dutch, *John Cheese* was *Jan Kees*. And that became *Yankees* when the English used the name to describe the Dutch settlers they were feuding with in Connecticut.

However the word began, it originally meant a person from Connecticut. Then it meant someone from New England; then someone from the North. Today, to many foreigners, a Yankee is any American.

sailors went down around South America and across the Pacific. Salem ships took the route across the Atlantic and around Africa. It didn't matter much; it always seemed to take two or three years to complete the voyage.

Captain Prince and Bowditch went to the Philippine Islands and back in 14 months. Now that was incredible! Captain Prince was able to sail his ship fast because Bowditch knew exactly where they were—he knew how to figure out latitude and longitude precisely and he was always right. Other ships had to wait for good weather, so their captains could see clearly to avoid dangerous rocks and currents.

Nathaniel wasn't content. He wanted his system to be easy enough for most sailors to use, so he simplified it and taught it to all the sailors on Prince's ship. Every one of them learned so well that each later became a captain himself. Bowditch printed his system in a book he called *The New American Practical Navigator*. That book

Seeing the Elephant

The first elephant ever seen in the United States came on the ship *America* in 1796. Captain Jacob Crowninshield paid $450 for the elephant on the island of Mauritius and, back home, charged admission to anyone who wanted to see him. The elephant "took bread out of the pockets of the spectators" and was a hit. Eventually the captain sold him for $10,000.

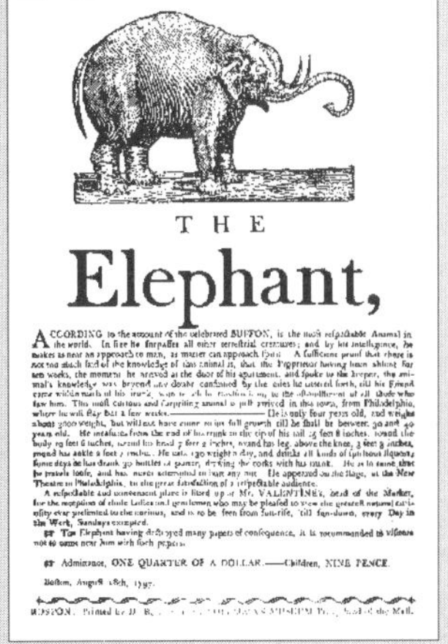

became one of the world's best sellers. Sailors all over the world used it way into the 20th century. It is still used by those who sail by the stars rather than with radar or other electronic equipment.

Nathaniel Bowditch became known as the "arithmetic sailor." Harvard named him an honorary doctor, and scientific societies in London and Berlin honored him. He wrote many books—on comets and meteors and mathematics—and he founded an insurance company. He made people in other lands respect the new country and the ability of its thinkers. When he died, in 1838, he was still a modest, thoughtful man. If ever you visit Salem, Massachusetts, look for a statue of Nathaniel Bowditch. On it you will see these words:

> *As long as ships shall sail, the needle point to the North, and the stars go through their wonted courses in the heavens, the name of Doctor Bowditch will be revered.*

In the War of 1812 Salem's merchants built a frigate, probably very like the *Philadelphia* (below), to help out the struggling U.S. navy. Their sons learned their fathers' trade as they played on the docks.

15 Thar She Blows

Joseph Palmer of Boston advertised his spermaceti candles bilingually—in French as well as English. Why did he do that?

Think of the most dangerous, scary amusement-park ride you have ever heard about. Whatever it is, it is tame compared to a Nantucket sleigh ride. That sleigh ride was probably the wildest ride the world has ever known. Some men rode it willingly. Some said they looked forward to it. Some did it because they got paid for the ride. Those men were very brave, or very foolhardy, or maybe both.

What is a Nantucket sleigh ride? Well, it has nothing to do with snow and everything to do with whaling. Hold on, I'll explain. But first, let's talk about the biggest creatures ever hunted—whales. Think about an extra-large school bus. That's about the size of a sperm whale. It can weigh as much as 60 tons. (Blue whales are even larger—they can be 90 feet long.) Whales are animals: warm-blooded animals that nurse their young—and swim like fish. But, unlike fish, they need to breathe air, and that is why they come to the water's surface regularly.

The sperm whale is a ferocious fighter with a huge jaw full of teeth. Sperm whales migrate great distances, although they prefer the warm waters near the equator.

Right, or *baleen* (buh-LEEN), whales have big mouths and tiny throats. Instead of teeth, they have long, plastic-like strips of baleen. The baleen lets water flow out of the whale's mouth, but holds tiny plants, fish, and animals—called *plankton*—inside. Right whales are pleasant creatures, usually found in cool waters.

Right whales got their name because they yielded a lot of oil and baleen, which made them, economically, the "right" whales to kill. Before the whalers got to work we think there were about 100,000 right whales in the oceans. Today there are probably 2,000 of them. (What percentage of the original total is that?)

When the cry goes up—*Thar she blows!*—the men pile into the sleek, graceful whaleboats. The boatsteerer, in the bow, his first iron (harpoon) ready, plunges it into the whale's flank.

The boatsteerer has driven in a second harpoon. Now the crew must back off as quickly as possible to be ready for whatever the whale will do next: attack, sound (dive), or flee.

Kerosene produces a better flame and a clearer light than whale oil—but it was hard to extract kerosene from its source—petroleum. When a reasonable extraction method was perfected, and when, in 1859, petroleum was discovered in Pennsylvania, it was the beginning of the end for the whaling industry.

Carved whale ivory is called ***scrimshaw.***

The Puritans' English charter granted them "all royal fishes, whales, baleen, sturgeons and other fishes."

All whales have a thick layer of fat, called blubber, under their skin. That blubber can be melted down and turned into oil. If you dip a wick in whale oil and light the wick, you can read by the light it gives. During the first half of the 19th century (until kerosene won the day), whale oil lit American homes and streets.

The sperm whale has a tank in its head filled with a liquid wax called *spermaceti* (spur-muh-SEE-tee). Spermaceti makes fine-quality candles. Some sperm whales have a bacterial substance in their intestines called *ambergris.* Ambergris is prized by perfume makers. Flexible whalebone—the baleen—can be used for things like buggy whips and women's corsets. The sperm whale's ivory teeth can be carved and made into decorative items. Because of all these uses, whales were very valuable creatures, especially in the days before electricity and plastics.

The Puritans knew that. They sometimes found a whale washed up on shore. They boiled its blubber and used the oil. Then the settlers on Nantucket Island noticed that the Indians went out in small boats and harpooned whales near the shore. So they did that too. In 1712, Christopher

Whalers spent days stuck below decks in bad weather. Since they saw no women for months, the ladies were favorite scrimshaw subjects.

This whale is sounding—diving straight down. The rope whips out so fast that the loggerhead it runs around has to be soaked with water to stop it bursting into flame.

The whale has reemerged and is attacking. The jaws of a sperm whale are strong enough to crush a boat. But now the men are close and can thrust the iron into lungs or heart.

Hussey, a Nantucket fishing captain, got carried out to sea by a storm. Hussey was lucky. He encountered a sperm whale, harpooned it, and towed it into shore. That was the beginning of the big-time American whaling industry.

At first, the whalers stuck close to New England, hauled the giant sea creatures to their home ports, and processed them on land. Then someone figured out that a hot fire could be safely kindled on a wooden ship in a brick oven surrounded by water. That meant the whale blubber was boiled in enormous iron kettles and turned into oil right on the whaler. Now ships could sail to distant oceans and stay at sea for three or four years—until their holds were filled with barrels of whale oil. That whale oil became a kind of liquid gold.

Near the Alaskan coast, Eskimos whale in long boats called *umiaks.* They waste none of the animal. Whale meat is a delicacy they enjoy. The Japanese, who are also fine whalers, find uses for the whole whale. They, too, feast on whale meat.

By 1846, more than 700 New England whaling ships were at sea, and New Bedford, Massachusetts, had become the capital of the whaling industry. The best whaling grounds were found in the Pacific Ocean, so that was where the whalers went.

For a boy from New England, to be a whaler might mean seeing the Arctic, the Antarctic, New Zealand, Japan, Guam, or the Hawaiian Islands. When

Taking off the blubber is called *cutting in.* The whale is lashed to the ship and strips of blubber, yards long and a foot wide, are sliced off with razor-sharp cutting spades.

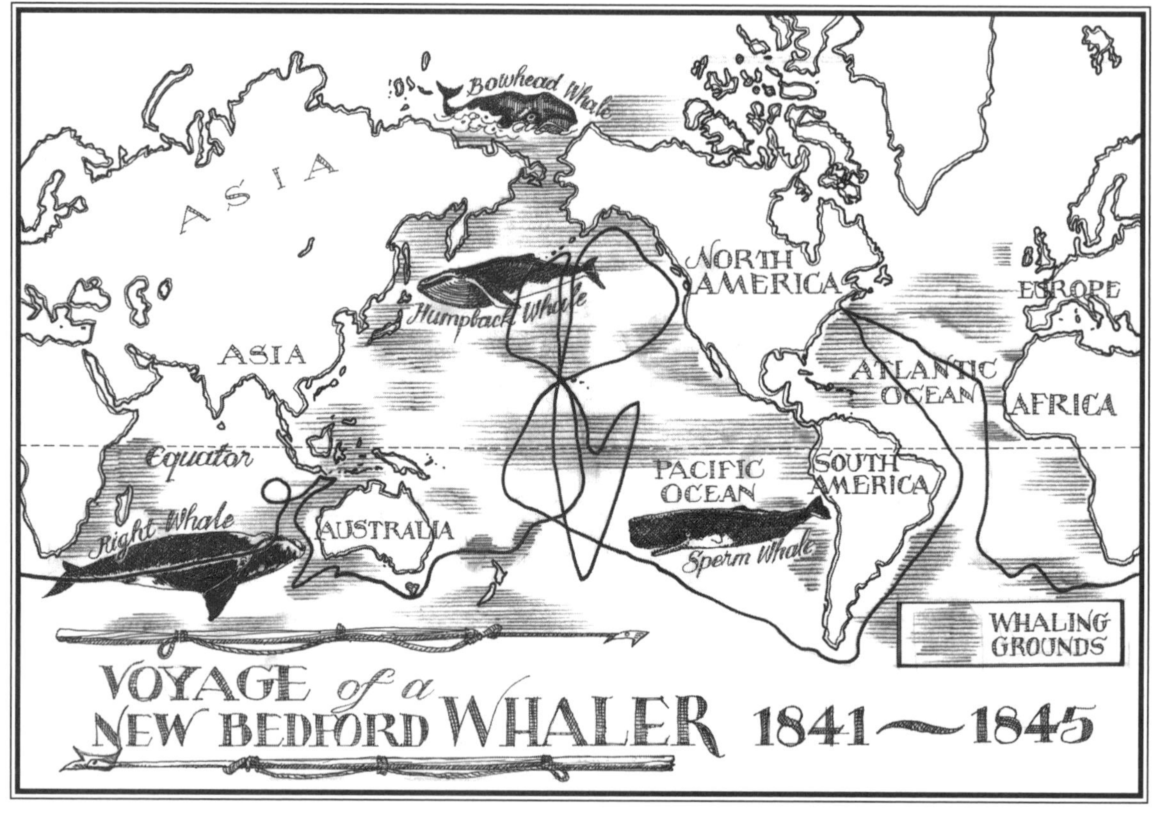

The whaleboat—like an Indian canoe—is light and speedy and said to be "the best small boat ever made." The mother ship, the three-masted whaler, has square sails and a broad deck.

those boys came home, it was with tales of adventure they would tell for the rest of their lives.

The officers of the whaler were usually, but not always, New Englanders of British descent. The crew was likely to be a mix of daring seamen: Native Americans, Portuguese, blacks, and Pacific islanders. Sometimes women and children were on board the ship. It was not unusual for the captain to have his family with him. The crew liked the idea. Food and conditions were usually better with a woman along.

How do you catch a whale? First you have to find one. Sometimes whalers go for weeks without spotting the towering waterspout that means a giant creature has come from the deep to breathe.

There you are, lazy and bored, when all of a sudden the lookout high on the ship's mast calls out, *"Thar she blows."* Then you see, off in the distance, the goliaths of the ocean.

Into a small whaling boat you jump—six men to a boat. The mate, in back, calls out, "Roar and pull, my thunderbolts! Lay me on, lay me on!" You pull on the oar as hard as you can. Then, as you approach the behemoth, all becomes quiet. The mate whispers; you row silently. The harpooner, standing in the front of the whale boat, is poised and ready. You need to get very close; that iron harpoon cannot be thrown far.

There it goes. It is a hit, deep in the whale's back—blood spurts high. A long, coiled rope is attached to the harpoon and anchored in your boat. Now the enraged whale takes off, trying to rid itself of the awful knife in its back. The rope plays out, like any fishing line, but so fast someone must throw water on it to keep it from burning.

The whale seems to be flying over and through the waves. You hang on and hold your breath. Now he is diving. If he goes too deep your boat will capsize. If he decides to turn around and attack he can easily destroy your boat. Some sperm whales have been known to

You can tell the age of a whale by checking the wax plug in its ear canal. Cut the plug and you'll see a layered pattern. Like tree rings, the layers show the passage of time.

A ***behemoth*** (bih-HE-muth) is a giant animal, like a whale or a hippopotamus. It's a word from the Bible's Book of Job.

From a Whaling Wife's Journal

In 1858, tiny Eliza Azelia Williams sailed with her husband Thomas William Williams, captain of the whaler Florida, *from New Bedford, Massachusetts, to the Sea of Okhotsk, north of Japan. When they returned home three years later they had two new shipmates—babies, both born at sea. Many years later, a son of one of those babies published Eliza Williams's journal. Here is an entry from that journal:*

September 29th. It is an unpleasant morning; the rain comes down in torrents. My husband has called me on deck to see the whale. It is in the water but fast to the ship; it is a queer looking fish....I can not say that I think there is much beauty to them; there is not much form, but a mass of flesh. Their flukes and fins are handsome. They are about a mouse color. Now they have...commenced cutting him in, and I can tell better how he looks. It looks like a monster to me and not a calf; it is quite long, and very shiny; there does not seem to be much form to the head; it is rather flat; a large mouth, quite small eyes. I thought they had no ears as I could not see any but my husband tells me that they have small holes about as big around as a knitting needle; he says that their hearing is very acute, and they have to go along as still as possible for fear the whale will hear the boats...the officers seem to understand exactly where to commence cutting.... They first take the blubber off with spades with very long handles; they are quite sharp, and they cut places and peel it off in great strips. It looks like very thick fat pork, it is quite white. They do not save any of the body but the outside, the blubber; but the head is the greatest curiosity.

A whaling wife watched from the balcony on her roof for her husband's return from the sea. Many men did not come back, and so the balconies became known as "widow's walks."

Prince Boston and Family

Prince Boston was a slave who lived on Nantucket Island in the days before the Revolutionary War. His owner, William Swain, sent him to sea on a whaling ship. Boston, the slave, was expected to do the whaling, and Swain, his owner, expected to collect the wages. But William Rotch, who was captain of the whaling ship, didn't think that was fair. Besides, he hated slavery. So he paid Prince Boston his own wages. Then William Swain died and John Swain inherited Boston. John Swain said Boston's earnings belonged to him. Swain went to court to try to get the money from his slave. William Rotch went to court, too, as Prince Boston's lawyer. When people in Massachusetts heard the news, they were so upset with Swain that he must have been embarrassed; he dropped the case. Soon after, in 1780, the Massachusetts Bill of Rights asserted that "all men are born free and equal."

Nantucket port, the whaling industry capital until a sandbar got in the way of the big 19th-century ships.

The Bostons became a respected family on Nantucket. Prince's son, Seneca, married an Indian woman named Thankful. They had six children and named the first of them Freeborn. You can guess why. Their youngest child, Absalom, became a ship's captain. In 1845, Absalom Boston's daughter was turned away from the all-white high school. Boston sued the town. It took two years, but Boston won that school integration case. His daughter went to high school.

In 1985, the International Whaling Commission (including 40 countries) banned commercial whaling. Since then some whales, like the sperm and gray whales, have made an amazing comeback. But pollution and overfishing (which eliminates much whale food) continue to menace the whale population. Blue, right, and humpback whales are still considered endangered species.

dive, come up under a boat, and toss it 20 feet into the air. None of those things happen on this day, but the whale does not tire easily. He pulls your boat and its six men around for almost an hour. This may be the world's riskiest profession. Finally the great mammal is exhausted. The mate spears him. The bloody deed is done. You row back to the ship dragging the huge carcass. This has been your first Nantucket sleigh ride.

But now there is no time to rest or think. The whale is strapped to the side of the ship. The blubber will be cut off in thick strips. It needs to be done quickly; sharks surround the ship. Yours isn't the only whale caught today. Four whale boats were lowered; four whales were harpooned: three were captured; one got away; all men are safe.

The crew works frantically. The blubber soon stinks in the tropical heat. Huge kettles are set to boiling. The blubber will be cut in pieces, boiled, skimmed, and strained. The process is called *trying*; it produces clean, clear whale oil.

The whale meat is wasted. Thrown to the sharks.

The whales' heads are brought on deck to remove the spermaceti and the ivory teeth. Oil and blood are everywhere. If you are not careful you can slip and fall overboard. If the crew is not careful the oil will catch fire and the wooden ship will be doomed. But all goes well. When you return to New Bedford you—and all the crew—will share the profits; you will have money in your pocket.

That is the reason for whaling. It is a business. It is all these things: brutal, destructive, useful, profitable, wasteful, and exciting. Whalers are as courageous as any men anywhere. But few people in the 19th century worry that a species is being destroyed. Like the buffalo, the very existence of the whale is being threatened.

This beautiful scrimshaw piece shows a whaler being outfitted at a San Francisco dock. Below: in 1820, the *Essex* of Nantucket, whaling in the Pacific 2,500 miles west of South America, was attacked, stove in, and sunk by a whale. The surviving crew got to Chile in their whaleboats three months later—after a terrible, grisly voyage.

16 A Boy from Japan

Japanese poetry is unrhymed. In brief poems called *haiku*, the poet captures a scene in nature; in *senryu* he comments on behavior and personality. (Usually senryu are about people—but I happened to pick an example about a horse and a bird.)

Little sparrow,
Mind, mind out of the way,
Mr. Horse is coming.

—KOBAYASHI ISSA

Spring rain:
Everything just grows
More beautiful.

—CHIYO-NI

It was an act of bravery that got John Manjiro his job as first mate. On a voyage to new whaling grounds off South America, the crew sighted a giant turtle. A ten-foot turtle meant food for all hands. When Manjiro harpooned it, the turtle struggled violently, so he jumped into the sea, got on the turtle's back, and plunged a knife into its neck.

Although John Manjiro returned to Japan, he never forgot his American friends and came back to visit in 1870.

Nakahama Manjiro was born in feudal Japan in 1827. In a feudal society, people are divided into classes: lords, knights, middle people, and peasants. Peasants are not allowed to own land; they have to work for a lord. In a feudal land there is almost no opportunity to rise from one class to another. It isn't as bad as a slave-owning society, but it isn't fair at all.

Much of Europe had given up feudalism by the 19th century, but Russia hadn't. Russia had serfs. Serfs were peasant workers bound to a landlord. If the landlord was horrible the serfs couldn't move or complain. Sometimes a landowner sold his serfs to another landowner. Then the serfs had to move. (In America, slave children could be sold away from their parents. That didn't happen in Russia. Otherwise, slavery and serfdom were not very different.)

Japan had its own version of feudalism; it was complex, orderly, and rigid. The emperor and the *shogun*—a military governor—were at the top of society, followed by the *daimyos* (lords), the *samurai* (knights), and, in descending levels, the rest of the people. In Japan, everything—the clothes you wore, the house you lived in, the food you ate, the job you did—was determined by the rank of your parents. If your parents were lower class, so were you.

Nakahama Manjiro was lower class. Besides that, his father had died when he was nine, so Manjiro had to support the family by fishing. One day, when he was 14, he was fishing with four friends. They didn't expect to stay out for long, but a storm came up and washed the oars out of their boat. There was nothing they could do. Actually they were lucky. They could have drifted forever. But after seven days, the boat

When he got back to Japan, Manjiro drew pictures of what he had seen. Then another artist took his drawing and made it into this print of the stern of Captain Whitfield's *John Howland.*

landed on an island. It was a deserted island. They were teenagers, on their own. Does it sound adventurous? They didn't think so. They lived in a cave, drank rainwater, and ate shellfish, seaweed, and birds. They almost starved. Six months went by. Then a ship came into view.

Men from the ship rowed a small boat ashore. Manjiro and his friends saw the men and were scared. They had never seen men with beards and white skin. They had never seen men with beards and black skin. There were some of each in the rowboat. But the Japanese boys didn't want to stay a minute more on the island. They shouted out to the men. They were American sailors, looking for turtles to make turtle soup. Their ship was a whaling ship, the *John Howland*, from New Bedford, Massachusetts. Captain William H. Whitfield took the castaways aboard.

Even if Captain Whitfield had wanted to, he couldn't have taken them back to Japan. Japan had closed its doors to all foreigners. No one but the Japanese was allowed on the islands. Any Japanese who had been in a foreign country risked death by returning to Japan. The country had closed itself off from the rest of the world. There were some advantages to that. Japan had lived in peace for more than 200 years. The Japanese had developed an artistic, refined culture. But there was no freedom in Japan. Some Japanese longed to travel.

An American in Japan

Ranald MacDonald's father was a Scotsman, an official of the fur-trading Hudson's Bay Company. His mother was a Chinook Indian. Her father, Chief Comcomly, greeted Lewis and Clark when they arrived on the Oregon coast. It was his Chinook blood that made Ranald dream of a distant land—a land closed to outsiders. He was convinced that the Chinooks were related to the Japanese. However, the Japanese Decree of Exclusion said: *So long as the sun shall warm the earth, let no Christian dare to come to Japan. ...if he violate this command, [he] shall pay for it with his head.*

Ranald MacDonald was willing to risk his life to find out more about that mysterious island. So he went to sea on a whaler, got the captain to set him afloat in a small boat, capsized it, and swam ashore to Japan, pretending to be shipwrecked. He probably didn't fool the Japanese, but they must have been fascinated by this brave and adventurous stranger. (All who met him commented on his charm.) He was closely guarded, but treated kindly. And he taught English to 14 Japanese scholars. After a year, MacDonald was put on a U.S. Navy gunboat and sent home. All this happened in 1848 and 1849; Manjiro arrived back in Japan in 1851.

Processing a whale on board the *John Howland*. Cutting spades in hand, the crew is peeling off the blubber while the try pots bubble in the center of the ship.

Some just wanted to do as they wished.

Captain Whitfield didn't attempt to return the boys. He was busy trying to catch whales. So he fed the youngsters and taught them the whaling trade. After five months at sea, the *John Howland* pulled into Hawaii. Four of the boys left the ship and went to live with American missionaries. (A missionary is a religious worker.) The captain invited Manjiro to Massachusetts to live with his family.

That's what Manjiro did. He was now called John Mung. In Japan, because he was a peasant, Manjiro had never been allowed to read and write. In Fairhaven, Massachusetts, John Mung went to school and learned to read and write English. In America he

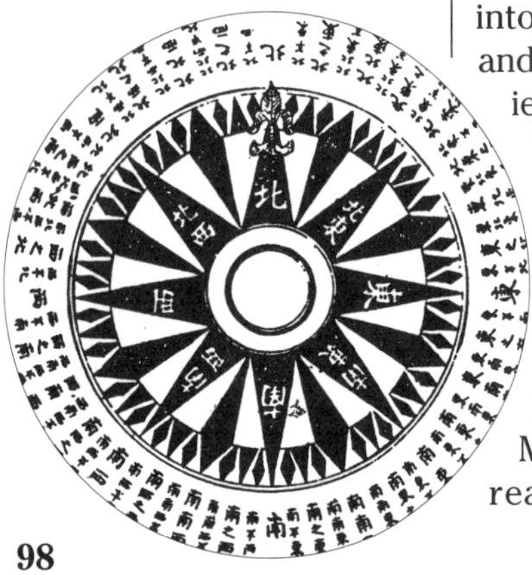

Manjiro in American sailor's gear. Manjiro translated Nathaniel Bowditch's *Practical American Navigation* into Japanese; on the left is his illustration of a compass.

learned to ride horseback. That was something else peasants couldn't do in feudal Japan.

Manjiro was a bright boy, and he learned quickly. He bought a copy of Nathaniel Bowditch's famous book on navigation. He studied it carefully. In America opportunity was open to people with talent. Manjiro had talent. He went to sea on a whaling ship and soon was a first mate. But he was homesick. He wanted to see his mother. He decided to risk death and return to Japan. First he went to California, to the goldfields. In two months he earned almost $600. He took his new wealth and sailed for Japan; he had much to tell the Japanese. But in feudal societies new ideas are often treated with suspicion.

For six months Manjiro was kept under guard and questioned by one daimyo after another. After being free in America, he didn't like it one bit. The Japanese officials didn't know what to do with Manjiro. Some were fascinated by what he had to say. Others wanted to keep him in jail. Then, in 1853, something astounding happened.

No Foreigners

Captain Whitfield took Manjiro to his church in Fairhaven. But when the deacon said that Manjiro couldn't sit alongside white people, the captain left that church. Then he joined a Unitarian congregation. Warren Delano was an important member of that church. Years later, Delano's grandson, Franklin Delano Roosevelt, became president and wrote to Manjiro's eldest son, inviting him to visit the White House.

Commodore Matthew Perry, of the United States Navy, sailed to Japan with four ships and 560 men. Two of the ships were steam-powered: they had neither sails nor oars. The Japanese wondered if they were driven by magic! Perry brought a letter from President Millard Fillmore to the emperor of Japan. It was sealed with pure gold. The president proposed in his letter "that the United States and Japan should live in friendship." He also said, "I am desirous that our two countries should trade with each other."

Commodore Perry and his staff arrive at the emperor's tent to negotiate treaty terms.

The Japanese leaders weren't interested. To them, the Americans seemed like barbarians. They asked Perry to leave. Perry wouldn't go. He wanted the Japanese to agree to sell coal and water and other supplies to whaling ships. He wanted them to agree to treat shipwrecked sailors courteously. He wanted Japan to agree to trade with the United States.

Commodore Perry left his letter. He gave the Japanese time to think about it. The shogun's advisers were divided: some wanted to learn about steamships and western technology. Others believed that western

Matthew Perry (above) had a brother who was also famous: Admiral Oliver Perry, the naval hero of the War of 1812. Below is a Japanese depiction of foreign ships in Nagasaki harbor.

ideas would destroy Japan's traditional culture. The shogun sent for Manjiro. No one knows what he said, but some historians believe that he helped make the Japanese look kindly on America.

When Perry returned, it was with a fleet of ships. He had studied Japanese ceremonial ways. He brought gifts, including: a telescope, a stove, two of Samuel F. B. Morse's telegraphs, five rifles, seven clocks, eight yards of scarlet cloth, a history of the United States, a copy of John James Audubon's picture book, *Birds of America*, a copper surf-boat, and a miniature railroad. The Japanese were wild about the little train. Everyone wanted to ride on it. "It was a spectacle," wrote a naval officer, "to behold a dignified mandarin whirling around...at the rate of twenty miles an hour...grinning with intense interest."

Perry's bands played loud music. He had his ships blast their cannons. He flew the Japanese flag on his ship next to the American flag. Then he told the Japanese that if they didn't sign a trade agreement with the United States, England or France would make them do it. He was successful. Japan agreed to end its isolation.

In 1860, when the first Japanese mission to America sailed for San Francisco, John Manjiro went with them. But he was not included in the official group that went on to Washington. He wasn't able to escape the handicap of his humble birth. Still, he wasn't a poor fisherman anymore. He was now an English teacher, and some of his students became important Japanese leaders. They were influenced by Manjiro's attitude to the West.

In 1868, a 15-year-old boy became emperor. He was Emperor Meiji, and his reign took Japan from feudal times into the modern world. Today Japan is a free, democratic nation; many Americans come from a Japanese background. John Manjiro would have liked that.

17 Cities and Progress

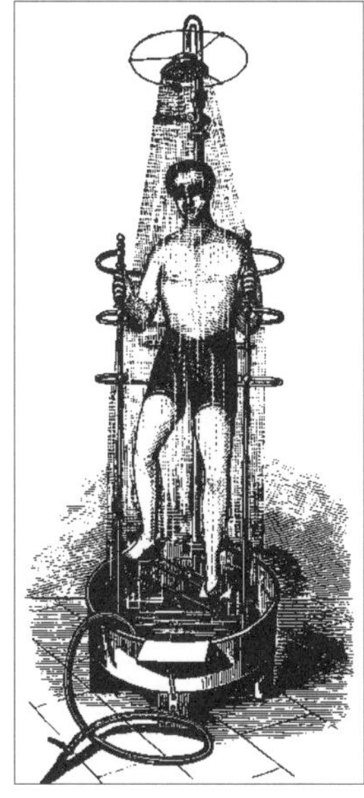

Indoor plumbing brought such novelties as this pedal-driven shower. It was advised only for those in the best of health.

It was a head-over-heels affair. The whole country was caught in its web. It was technology that had captured us. We Americans, in the 19th century, became fascinated with machines and scientific advances. We watched as they changed our old ways, and, mostly, we liked what was happening.

We fell in love with speed—with locomotives and steamboats and clipper ships.

We fell in love with American inventions—with John Deere's steel plow, Cyrus McCormick's reaper, Elias Howe's sewing machine, Charles Goodyear's vulcanized rubber, and Samuel Morse's electric telegraph.

We fell in love with indoor plumbing.

And many of us fell in love with city life.

Back at the end of the 18th century—in 1790, when the first census was taken—95 percent of all Americans lived on farms. By 1820, 93 percent of Americans lived on farms or in rural villages. By 1850, it was 85 percent. Slowly, we were on the way to becoming an urban nation. Where do you live? Where do most Americans live today?

In 1790, only two cities—New York and Philadelphia—had 20,000 or more people. By 1860 there were 43 American cities of at least that size, and another 300 cities with more than 5,000 inhabitants.

Technology (tek-NOL-uh-jee) is the use of scientific ideas for practical purposes.

Moon Madness

In 1835 the *New York Sun* put its tongue in its cheek and decided to see if its readers could be fooled. The newspaper reported that an English astronomer, Sir John Herschel, had discovered that there was life on the moon. Quoting an article in the *Journal of Science*, published in Edinburgh, Scotland, the *Sun* ran Sir John's description of moon creatures with furry bodies and batwings. Other newspapers copied the story—and, yes, people were fooled. Scientists from Yale University came to New York to read Sir John's account. Finally, the *Sun* revealed that there was no Sir John Herschel and no *Journal of Science*. But by that time the *Sun*'s circulation had leapt to 19,360, the largest in the world.

This is the city and I am one of the citizens,
Whatever interests the rest interests me, politics, wars, markets, newspapers, schools,
The mayor and councils, banks, tariffs, steamships, factories, stocks, stores, real estate and personal estate.

—Walt Whitman, from "Song of Myself"

New York, according to the poet Walt Whitman, was "million-footed." But that was only when New Yorkers hopped on one foot. A million people lived in New York. Philadelphia had 500,000. Cincinnati, known as the "Queen City of the West," had 160,000. New Orleans, "Queen of the South," had 169,000.

What were all those people doing in cities?

Most were working—making things, or teaching, or selling, or preaching. Many were new immigrants who stayed only a short time before they went off to look for opportunity elsewhere. They moved and then moved again; for Americans were a restless people, always searching for a better life.

We Americans became famous all over the world for "know-how." People in this country learned to use their hands and heads to make things, and make them well. In the 19th century the United States was becoming an industrial nation. Industry was helping cities grow.

If you've never been in a city before (and many 19th-century Americans had not), it's a mighty exciting place. Why, just look at that building over there—it must be six stories tall. You can get a stiff neck looking up that high! Do you think there is any danger it will topple over? It's the handsomest building in town—and it's a hotel. Step inside and you'll be in a big lobby, where trees grow in pots.

They say that some American hotels even have vertical railroads inside. (City folk call those vertical railroads *elevators.*)

Boston's Tremont House, built in 1829, is the finest hotel in the country. It has eight "bathing rooms" in the basement. Cold water comes into the baths from rooftop tanks that collect rain. Now listen to this: the Tremont House basement has eight water closets (someday they'll be called toilets) for the 200 to 300 guests.

New York's Astor House Hotel, built in 1836, has running water above the first floor. (It is the first public building to try that.)

And in New York, A. T. Stewart's Department Store, six stories high and built of white marble, has 400 clerks and big glass windows to show the splendid merchandise sold inside. It is a new idea—a store that sells almost everything you could want. People visit New York just to see Stewart's store.

Cities all over the land are beginning to have their own versions of Stewart's and the Astor House.

One new idea led to another. Stewart's department store (top) was one of the first cast-iron buildings in America, and the invention of rolled sheet glass made its big windows possible.

New York's bustling waterfront at South Street. You can visit the restored seaport there today, tall ships and all.

Of course, every city has a newspaper; some have five or six or more. Newspapers are changing the nation. In 1830 there are 1,000 newspapers in America. Mostly they are sold by yearly subscription—at about $8 a year. That's a lot of cash; most people can't afford to read a paper. That is, until 1833, when Benjamin Day publishes a newspaper, the *New York Sun*, that sells for a penny. Two years later he is selling nearly 20,000 copies of the *Sun* every day—and lots of advertising too. James Gordon, newly arrived from Scotland, soon follows with the even more successful *New York Herald*, and Horace Greeley with the *New York Tribune*.

By 1840 the nation has 138 daily papers and 1,141 weeklies. Readership is growing at an even faster rate than population. Now, everyone seems to be a newspaper reader.

Some people complain that the penny papers are vulgar and trashy, but there are newspapers for every taste. They are wildly popular and help spread information and democracy, too. (Everyone can afford to read them, not just the rich.) They are part of the excitement of city life.

Farm life is quiet; city life is not. In America you can have your choice.

But building cities quickly means there is little time for planning—and that sometimes leads to trouble. Have you ever heard adults talk

Horace Greeley of the *New York Tribune* wrote the famous words "Go west, young man." He ran for president in 1872 but lost to Ulysses S. Grant.

"The typical tenement house ...has no light from the open air, no ventilation....On its steps play the pale, unhealthy children who...still swarm in these horrible dwellings."

about the "good old days"? Well, they weren't as good as some people think.

Crime was an awful problem in the 19th century. Most American cities didn't have uniformed police, but they did have street gangs, pickpockets, and robbers.

There just weren't enough rooms for all those people who were moving in; the cities were overcrowded—and dirty. Boston had a good sewage system, but most cities didn't.

Waste water from baths and sinks went into open street gutters. Pigs roamed city streets, and so did rats. Chicago had almost no indoor plumbing (until after 1860). No one understood the importance of cleanliness, so disease was a terrible problem. Whole neighborhoods were sometimes wiped out by cholera and other deadly diseases.

And then there was fire. Most city buildings were wooden and built close together. If a fire was ignited—*whoosh*—a city could burn down.

It was December of 1835, and cold; New York's fire hydrants were all frozen. So the city's firemen could do nothing but watch when a fire got

The traffic on Broadway in New York was terrifying—and the dirt, garbage, and horse manure in the street almost as distressing—so Mr. Genin, the hatter, built this bridge in 1852 at his own expense.

started. Soon most of New York was ablaze—700 buildings turned to ash. That happened in more than one city.

For all that, America's cities were full of good surprises. An Englishwoman visiting America in 1834 (her name was Harriet Martineau) was amazed to hear a concert of Mozart's music in Cincinnati, Ohio. A few years earlier, she was told, the only sounds there were "the bellow and growl of wild beasts."

New York City's terrible fire of 1835.

Most Americans were sure they were in the best possible place in all the world. They knew that if the farm or city where they lived didn't work out, they could move on: to a new city sprouting in the wilderness, or to a new frontier. America's land was so vast that much of it was still unmapped. That added to its allure. It seemed limitless, and so did the opportunities it offered. Those new cities were fascinating, but it was land of their own that most Americans wanted.

The Richest Man in America

John Jacob Astor arrived in Baltimore from Germany with seven flutes. If he could sell them, he thought, they would get him started in the New World. They did. When he died, 64 years later (in 1848), he was the richest man in America and he owned a whole lot of Manhattan island (which is the center of New York City). Astor started by selling furs and musical instruments, and soon had a shop in New York City. Besides that, he was buying and selling goods to the Indians at frontier outposts. He began importing guns, ammunition, and wool. That wasn't all: he acquired a fleet of ships and sent some of them off to China. He controlled most of the Pacific Coast fur trade. He created a trade network linking Britain, New England, the Pacific Coast, and East Asia—and he did it long before others thought globally. Then Astor began buying New York real estate. He was soon called "the landlord of New York," and he became super-rich. He said he was sorry that he hadn't bought all of New York City.

What was he like? He was tight. It took the artist John James Audubon six trips to see him before Astor would pay the $1,000 he had promised for a copy of Audubon's masterpiece, *Birds of America*. He told Audubon he didn't have any cash. But he did found some libraries, and he gave money to colleges and a few other worthy causes.

18 A Land of Movers

Even little children carried water and gathered buffalo chips. And older kids regularly helped with the heavy chores.

Imagine: it is 1829. Your name is Jacob and you live in Indiana. You like it here. The hilly green land is covered with forests and laced with creeks and rivers. Duck and geese fill the sky. Sixteen years ago your father was living in Germany. Then he emigrated to Pennsylvania, met a Scotch-Irish girl, married, and set out to pursue happiness. The year you were born—1816—your family moved here, just north of the Ohio River. They believe they made a wise decision: that very year Indiana was named the 19th state.

You don't remember your first year in Indiana, but you've heard stories of how you lived in a dirt-floored hut while your parents cut down trees and began to farm and to make life better for themselves. Your father built a one-room cabin. There, your mother cooked in a big pot hanging in the fireplace. She spun, wove, and sewed all the family clothes.

But your father wasn't much of a farmer. Besides, he likes to be around people. So your family and some others started a town. Now you live in a neat, four-room wood-frame house. In your town there are: 84 houses, a church, a brick school, eight stores, two taverns, a newspaper office, a courthouse, and a library with 300 books. There is not one divorced person. Some married people you know don't like each other much, and some men in town are said to have deserted their families, but divorce is rare. Alcoholism is not. Your town has a number of alcoholics who are public nuisances, and others who drink quietly.

This was the Northwest Territory. According to the law there are

Jacob eats mostly salt pork and corn. Sometimes he has cabbage, beets, or turnips, but, otherwise, few vegetables or fruits. His parents worry about diseases like tuberculosis and diphtheria, but people don't know what to do to protect themselves from those dread afflictions. And no one has told them that there are health reasons for washing your hands or brushing your teeth. Jacob can expect to live to be about forty. If he lives longer, he will have passed the national average.

supposed to be free schools and no slavery. But a few people do own slaves. There are public schools, but not enough of them. Your parents pay a schoolmaster and you go to his school. All the students sit together in one room. The older boys and girls help teach the younger ones. You are one of the teacher's aides. You have become a good reader and a fine penman. You have a new, steel-point pen—instead of a homemade quill pen—and you make ink from wild berries, vinegar, and salt. When school is out you take the cows to pasture, chop wood, and haul water. Sometimes you catch fish for the family dinner. As often as you can, you play marbles, and bat and ball, and football with a washed pig's bladder you have blown up.

On Saturdays, you work in your father's store. He is called by that new word, *businessman*. He farms, and he sells nails and lumber in his general store. He also sells land. If the town grows he will become an important man; perhaps he will enter politics.

Five years ago you had a friend, Ohiyesa, a Shawnee Indian boy. Now he is gone. Ohiyesa told you of happy days when he was small. He remembered big gatherings when all his relatives came together and feasted on buffalo steaks, and maple sugar, and corn stew. Ohiyesa told you of Tecumseh, a great Indian leader who was killed in battle. He told you how he and his family wept. That was the end of the Indian

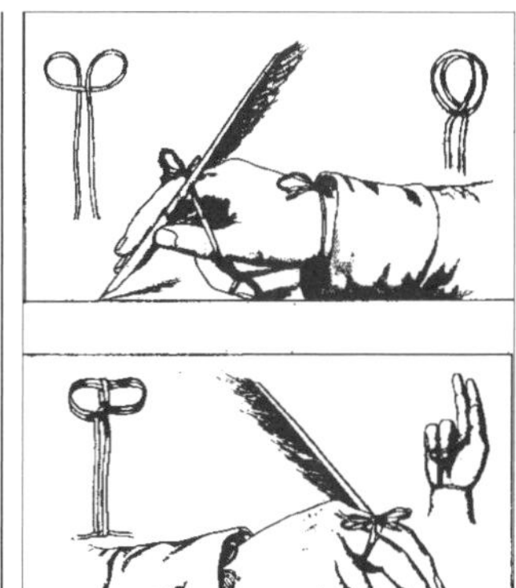

For his handwriting manual, *Practical Penmanship* (1832), Benjamin Franklin Foster devised a way of training the hand to move fluently by tying fingers and wrist into place.

William Forster, a Quaker minister living in Indiana, wrote in 1821: "I am sorry to say there are many slaves in the town.... I wish some active benevolent people could induce every person of color to remove away from the river, as it gives wicked, unprincipled wretches the opportunity to get them into a boat, and carry them off to Orleans or Missouri, where they still fetch a high price."

A Kansas family and the wagon that got them there. "It might seem a cheerless life," wrote one woman, "but there were many compensations."

From the
Indianapolis Journal, **September 17, 1829:**
"The tide of emigration seems as usual to be flowing to the west. Our streets begin already to be crowded with families, and wagons, and stock. ...Counties which three or four years ago, were but a trackless wilderness, contain now five, six, and seven hundred voters."

celebrations. It was the end of the Indian confederation. Tecumseh had said the Native Americans and the newcomers could live peacefully in the same country. But he was wrong. They can't seem to do it. The Indian way was doomed when the settlers began coming. Every lot your father sells, every tree cut down by a farmer, changes the land. Your way of life has cost Ohiyesa his home. It isn't your fault, is it?

You have another friend. Her name is Ida. Last year Ida was a slave in Mississippi. Then her family heard that she was to be sold; they might never see her again. When they learned the news, they made a decision. They would run away to freedom. They had listened to stories of slaves who went north following the tail of the Big Dipper in the sky. They had sung songs of freedom and dreamed of a world of fairness. They were ready to take a chance.

So they followed the stars and the big river, escaped the fearful dogs sent to find them, and made it to Indiana. Ida needs to be careful. Blacks are sometimes kidnapped and enslaved. She goes to a Quaker school. There are thousands of Quakers in Indiana. They have encouraged blacks to live in or near their settlements. George Washington's profile hangs in Ida's house, and she knows some of Thomas Jefferson's words by heart. She has made you proud of our nation and its goals.

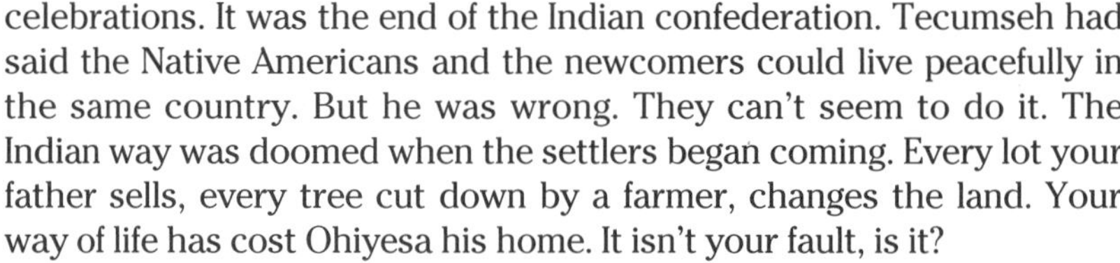

Left: a Kansas school equipped with a real blackboard. Most did not even have floors. Below: a sod-house family. The rain came through the roof, but the cow didn't.

Americans have already achieved a national identity, which is astounding when you consider that our backgrounds are so different. Someday some people will call this nation a "melting pot"; others will call it a stew. It doesn't matter. People are coming here from all over to add spice to the pot. Soon after they arrive they are all Americans.

Some of them come to Indiana, where you—Jacob—have almost forgotten Ohiyesa and your boyhood years. Time has passed, and you are getting on with your life. You have studied law and become a state senator.

Your father has built another new house. This one has a cast-iron stove in the kitchen and one of Dr. Benjamin Franklin's stoves in the parlor. Your

An apple bee (above), a quilting bee, corn huskings, barn raisings—all were ways of getting big tasks done while having fun at the same time. Many artists of the time, like this one, saw these occasions mainly as excuses for young men and women to get together.

mother cooks in a skillet on the kitchen stove. (The family diet still centers around pork and corn.) The stoves are efficient; they use only a small amount of wood or coal. Wood is becoming scarce.

Your mother has stopped spinning wool and flax. Factories are turning out cloth. She doesn't have one of those newfangled sewing machines, but she would like one. Why, they are said to stitch a seam in seconds.

But you don't live with your parents anymore. You have built a grand house of your own. You are married now, to Maureen Kelly, who came here from Ireland with her family to escape the Great Potato Famine. Your wife doesn't want to haul water from outside, so you put a water pump right in the kitchen. Inventors and dreamers are saying that someday there may be machines to wash clothing (instead of boiling, scrubbing, and wringing, all by hand), and water closets (toilets) inside people's houses (instead of going outside in the cold of night), and horseless carriages—but all that may just be wild imaginings.

Many Hands

The help of the neighbors was extended to many kinds of work, such as rolling up the logs in a clearing, grubbing out the underbrush, splitting rails, cutting logs for a house, and the like. When a gathering of men for such a purpose took place there was commonly some sort of mutual job laid out for the women, such as quilting, sewing, or spinning up a lot of thread for some poor neighbor. This would bring together a mixed party, and it was usually arranged that after supper there should be a dance or at least plays which would occupy a good part of the night....

The flax crops required a good deal of handling, in weeding, pulling, and dressing, and each of these processes was made the occasion of a joint gathering of boys and girls and a good time. As I look back now upon those times, I am puzzled to think how they managed to make such small and crowded houses serve for large parties.

—William Cooper Howells, from *Recollections of Life in Ohio, 1813 to 1840*

Prairie settlements became thriving towns (this is Topeka, Kansas), and the first pioneers watched as the next arrived, headed even farther west.

19 Workin' on the Railroad

"I like to see it lap the miles, / And lick the valleys up, / And stop to feed itself at tanks; / And then, prodigious, step / Around a pile of mountains...." What was the poet Emily Dickinson referring to?

You could ask anyone, anywhere in the United States, to tell you who was a symbol of everything good and noble about America—and the answer would always be the same. It was a man who represented honesty and duty and patriotism. A few people could still actually remember him. He was George Washington, and his birthday, along with the Fourth of July, was an important holiday.

So, if you had a big event to celebrate, February 22 was the perfect day for it. And when the first railroad train made it from the East Coast to the Mississippi River, it was an enormous event. They were still spiking down railroad ties just an hour before the train came whistling and hooting into view at Rock Island, Illinois, on George Washington's birthday in 1854. The train, the Chicago & Rock Island Line's Number 10, was covered with wreaths and red, white, and blue banners.

English writer Charles Dickens said: "There are no first and second class carriages as with us; but there is a gentlemen's car and a ladies' car...in the first, everybody smokes; in the second, nobody does."

Who thought up the idea of railway tracks? No one knows, but we do know that the ancient Greeks, Romans, and Assyrians understood that wheels roll easily on a smooth track. They cut grooves in stone slabs and rolled wagons along those tracks. Their vehicles are long gone, but some of those tracks can still be seen.

Competition with steamboats was not the only question raised by the construction of the Rock Island Line to Iowa. It fueled the arguments between the South and the North about slavery. You can read more about that in chapter 32.

CHICAGO AND ROCK ISLAND

AND

MISSISSIPPI AND MISSOURI RAIL ROADS.

Only one change between Chicago & Ft. Des Moines

THROUGH TO IOWA CITY IN 10½ HOURS,

Without Change of Cars or Baggage.

THE COMPLETION OF THE

MAMMOTH R. R. BRIDGE

At Rock Island enables the passenger, via this route, to make the transit from Illinois to Iowa without encountering delays and dangers of ferrying the Mississippi River in Winter.

Passengers will Notice this Fact.

That the CHICAGO & ROCK ISLAND RAIL ROAD, and its connections, are 27 miles nearer Fort Des Moines than any other route, and only one change, making 12 hours time in favor of this route. No other route can take them within 90 miles of Iowa City by Rail Road.

This route offers superior advantages to passengers going to CENTRAL AND WESTERN IOWA, KANSAS AND NEBRASKA, it being the SHORTEST, CHEAPEST, QUICKEST AND SAFEST, MORE RAIL ROAD, AND LESS STAGING, THAN ANY OTHER.

Companies going to the Territories can purchase or hire teams at Iowa City at moderate prices; and those wishing to settle in Iowa will find the most valuable lands in the vicinity of this, the Great Overland Route to the West, better timbered and watered than any other part of the State, and offered at moderate prices and easy payments.

THREE TRAINS LEAVE CHICAGO DAILY, AS FOLLOWS:

9.10 A. M. SUNDAYS EXCEPTED. DAY EXPRESS, for La Salle, Peoria, Rock Island, Iowa City [illegible]

2.30 P. M. SUNDAYS EXCEPTED. MAIL ACCOMMODATION, for Rock Island.

11.00 P. M. SATURDAYS EXCEPTED. NIGHT EXPRESS, for Peoria, Rock Island, Iowa City and Muscatine.

ALSO TWO DAILY TRAINS FOR St. LOUIS.

DAILY LINES OF STAGES LEAVE IOWA CITY FOR ALL THE PRINCIPAL TOWNS IN WESTERN IOWA.

ALL BAGGAGE RE-CHECKED AT CHICAGO

THROUGH TO IOWA CITY, AND TRANSPORTED FREE FROM CONNECTING ROUTES.

THROUGH TICKETS VIA THIS ROUTE

Can be purchased at all Eastern and South-Eastern Rail Road Offices, and at the Company's Offices in Chicago.

W. L. St. JOHN, General Ticket Agent.

JOHN F. TRACY, General Superintendent.

A. M. SMITH, TRAVELING AGENT.

CHICAGO JOURNAL STEAM PRINTING ESTABLISHMENT.

The locomotive pulled six passenger cars filled with people all waving flags and handkerchiefs. A second train, also decorated and filled with passengers, followed close behind. Musicians struck up "Hail Columbia." They were followed by speechmakers (including the governor), who weren't going to let this day pass without getting their words in. Then everyone sat down to eat at banquet tables set up in the brand-new train depot. Someone offered a toast "to the Press, the Telegraph, and the Steam Engine, the three levers which move the world of modern civilization." A man standing near the door signaled the train engineer, and the Iron Horse applauded with a shrieking blast from its whistle.

Not everyone was thrilled about the coming of the railroad. Rock Island was a steamboat town. So was Davenport, across the big river in Iowa. What would railroads do to steamboats?

You didn't have to be brilliant to figure out that the wheat and corn and hogs that were being shipped from the middle of the nation down the Mississippi and then around to East Coast ports could go by train—due east. If that happened, railroads would make the city of Chicago more important (it was a railroad center), and the Mississippi River cities—like St. Louis and New Orleans—less important.

No one was quite sure of all that in 1854, but they knew the railroad would bring big changes. Most people liked the idea of change; it seemed exciting. But not to those steamboaters.

The men who'd planned the train-depot dinner didn't care what the steamboat people thought. This was just the beginning of their dream. They intended to take the

In the decade between 1820 and 1830, 40 miles of track were built in the entire United States. By 1835, 1,000 miles of track had been laid, and more than 200 railroad lines were being planned or constructed. By 1850, 9,022 miles of railways were operating. And by the time the Civil War broke out, in 1861, the total track distance was three times as large as it had been 10 years before—31,246 miles.

A Mississippi steamboat in 1858, two years after the Rock Island Line was finished. The bridge, the cause of all the trouble, is visible in the background.

There were other reasons than fear of competition for worrying about the spread of the railroads. Boilers often blew up (on boats, too), killing many and inspiring this grim cartoon.

railroad right across the broad Mississippi. Then they wanted to go even farther. They planned to make tracks across the nation. They'd brought the railroad to the town of Rock Island because there was an island with that name nearby, in the middle of the river. They could build a bridge to the island and on to the western shore.

But as soon as construction began on the bridge, steamboat owners started to scream. They said the bridge was "unconstitutional, an obstruction to navigation, dangerous." They weren't the only ones concerned. The train tracks were following a northern route. Southerners wanted the first railroad across the continent to go through southern territory. They knew it would bring commerce and industry with it. They wanted that business for the South. So they went to court; but they lost. Construction continued, and, in 1856, the bridge was ready. A newspaperman was on board when the first train crossed the river. This is what he wrote:

> *Swiftly we sped along the iron track—Rock Island appeared in sight—the whistle sounded and the conductor cried out, "Passengers for Iowa keep their seats." There was a pause—a hush, as it were, preparatory to the fierceness of a tornado. The cars moved on—the bridge was reached—"We're on the bridge—see the mighty Mississippi rolling on beneath"—and all eyes were fastened on the mighty parapets of the magnificent bridge, over which we glided in solemn silence. A few minutes and the suspended breath was let loose. "We're over," was the cry. "We have crossed the Mississippi in a railroad car."*

Back East, an article in the *Philadelphia Bulletin* said, "Civilization took a railroad trip across the Mississsippi."

But the steamboaters didn't care what newspapers said or what most people believed; they were angry. Still, no one expected what happened two weeks later, when alarm bells woke the citizens of Rock Island and Davenport. The wooden parts of the bridge were on fire! The *Effie Afton*, a boat from New Orleans, had had an accident (or so it was said). The boat had run into the bridge. The collision had overturned the steamboat's stoves and set her afire. The fire had spread to the brand-new bridge.

Was it an accident, some people asked? That afternoon a banner went up on one steamboat. It said, MISSISSIPPI BRIDGE DESTROYED. LET ALL REJOICE.

The owner of the *Effie Afton* sued the bridge builders. He said the bridge got in the way of his boat. He said the bridge was a navigation hazard. The railroad men needed a lawyer. They asked around and got the best Illinois lawyer they could find. His name was Abraham Lincoln.

Lincoln, who was in his early 40s, was tall and lean, with a thick tangle of dark hair and no beard. He was a solemn-looking man, but known for his whimsical humor. He was also known to be the kind of lawyer who spent plenty of time preparing his cases. He did his homework carefully.

Lincoln walked out on what was left of the bridge, took a teenage boy with him, and then figured out the speed and direction of the water's current by timing the movement of twigs that his young friend dropped into the water.

The trial began and it was soon clear that this was North against South as well as railroad against steamboat. Whoever won would control the commerce of the growing Midwest. Abraham Lincoln sat quietly whittling on a stick, but when he stood up to question a witness he seemed to know exactly how big the bridge was, and how deep the water, and the exact size of the *Effie Afton*, and what its normal schedule was, and what the river currents were, and what the steamboat captain had to do to avoid the bridge. Besides all that, Lincoln kept saying that railroads had just as much right to cross over rivers as steamboats had to go up and down them.

The jurors couldn't reach a decision, but that was seen as a victory for the railroad people. And when the case was appealed—all the way to the Supreme Court—it was made clear: railroads had the right to bridge rivers.

The great steamboat era: top, outside, in the incredibly busy port of New Orleans in the 1880s; bottom, inside, stoking the boilers. But east-to-west commerce across the country was controlled by the railroads.

20 "She Wishes to Ornament Their Minds"

"The school was about 14 feet long with dirt floor, unplastered walls, two windows. . . ." "They had a few bare benches. . . .No maps, no pictures, no books but a Speller."

The first public school in Florida opened in Tallahassee in 1852.

If I tell you that in the first years of the 19th century, almost half the boys and girls in the United States never went to school, you might think it a backward time. But remember, when you read history you need to put yourself in a time capsule and zoom back and try to think as people did then. If you do, you will find that America's citizens thought the United States was the most exciting, progressive place in the whole world—and many people in other nations agreed with them.

A larger percentage of children went to school in the United States than in any European country. (The 1850 census showed that 56 of every 100 white children went to school—but only two of every 100 black children.) Many children who never went to school still learned to read. In those days, boys and girls often spent more time in church than in school. So some learned to read in Sunday school, and just about everyone read Bible stories or heard them read. Many parents taught their children at home.

Unfortunately, Jefferson's school plan wasn't even used in Virginia. It was turned down in the legislature. Most of the Virginia legislators were wealthy men, who sent their children to private academies or had tutors for them.

Before 1830, there were few free public schools outside New England. New England had a tradition of public schools. Massachusetts set up what may have been the first free public-school system in the world; that was way back in 1647. In other places, free schools were mostly for poor children. Wealthy children had private tutors or went to private academies. And, in many communities, parents got together,

hired a teacher, and built a schoolhouse.

The public-school idea began to catch on in the 1840s and 1850s. America was becoming more democratic—more men could vote—and the new voters began demanding schools for their children. The states started establishing school systems.

The Constitution didn't mention education. That wasn't because the Founders thought education unimportant; it was because they expected the states to control schooling. Thomas Jefferson believed the American experiment in government—that new idea called self-government—would only work in a country where every citizen was educated. After all, how can you vote and make decisions if you can't read?

This was a song of the 1830s: "Is it not a cruel fate a master thus to be, / Doom'd to teach such naughty boys, such blunder heads as these.... / letters telling, hard words spelling; / Pens a making, boys a shaking, / Reading writing scolding fighting, / Coaxing on the stubborn ones, pushing on the lazy."

"If a nation expects to be ignorant and free, in a state of civilization, it expects what never was and never will be," wrote Jefferson in a letter to a friend. He devised a plan for schools in Virginia. He hoped it would be a model for other states too. George Washington said, "a plan of universal education ought to be adopted in the United States." (In other words, he wanted a national school system.) And he also said, "Knowledge is, in every country, the surest basis of public happiness."

Sarah Pierce agreed with that idea. She started her own school, the Litchfield Female Academy, in the dining room of her Connecticut house. Most people thought girls should concentrate on needlework and music and painting—but not Sarah Pierce. In her school girls were taught grammar, reading, composition, history, philosophy, and logic—as well as needlework. Nancy Hale, who was one of her students, wrote to a friend that Miss Pierce did not "allow anyone to embroider without they attend to some study for she says she wishes to ornament their minds when they are with her." Her school existed for four decades (from 1792 to 1833) and attracted 3,000 students from 15 states, the territories, Canada, and Ireland. Harriet Beecher Stowe was a student at the Litchfield Female Academy (more about Harriet in Book Six of *A History of US*).

Tapping Reeve Law School

Students at the Litchfield Female Academy did more than study. They had a busy social life too. The well-respected Tapping Reeve Law School was nearby, and many girls married law students. (John Calhoun went to Tapping Reeve.) Charlotte Sheldon wrote in her diary that she "dressed and danced in the evening, had a pretty agreeable ball." The tuition (it covered schoolwork and needlework) was $5 a semester. Students boarded with local families for $2 a week.

Noah Webster, who was also born in Connecticut, believed that American children should read American schoolbooks, not the English books they were reading. So he wrote American schoolbooks and an American spelling book. Then he began writing an American dictionary. Webster put everyday words—even some slang words—in his dictionary. That was unusual. It took him 20 years to complete his two-volume *American Dictionary of the English Language*. It was published in 1828. (Do you have a Webster's dictionary in your house or school? Can you imagine writing a dictionary?)

Before Webster wrote his books, people spelled words any way that made sense to them. As you can imagine, words sometimes got confusing. Webster ended the confusion: he established standard spellings. His speller was soon the best-selling book in America—after the Bible. Because of Webster's speller, schools everywhere had spelling bees. They were very popular. And talk about democratic! A boy or girl from a little country school could beat students from the fanciest academies and win the title as the best speller in the state.

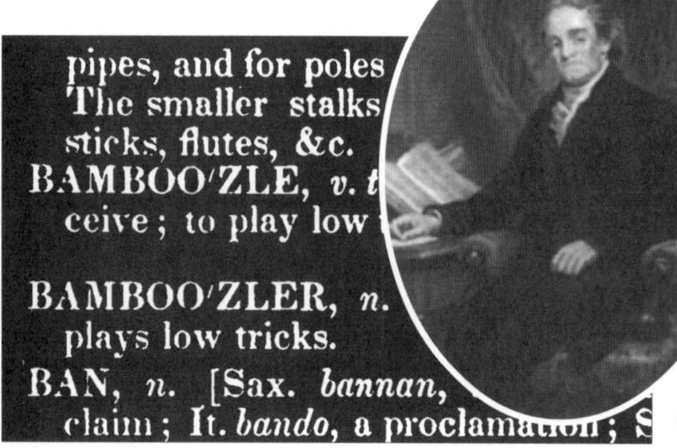
pipes, and for poles
The smaller stalks
sticks, flutes, &c.
BAMBOO'ZLE, *v. t*
ceive; to play low
BAMBOO'ZLER, *n.*
plays low tricks.
BAN, *n.* [Sax. *bannan*,
claim; It. *bando*, a proclamation;

Noah Webster (above) graduated from Yale College, and was a teacher, a lawyer, a journalist, a scientist, and a patriot. Many dictionaries today have the name *Webster* in their title. They are not Noah Webster's work. The name may be used by anyone, because book titles cannot be copyrighted. (What is copyright?)

But if a girl won the spelling bee or was the best student in her class, it didn't matter. She still couldn't go to college. That bothered a girl named Mary Lyon. She wanted to go to college.

Mary was unusual. One day, when she was little, her mother found her fiddling with an hourglass. "What are you doing?" she asked. Mary said she was trying to invent more time. Mary Lyon never seemed to have enough time to do all the things she wanted to do. Sometimes she studied 18

The first women's educators, like Mary Lyon (right), often had to reassure parents that their daughters were not learning subjects unsuited to their future role as housewives and mothers. For instance, chemistry was considered acceptable because its principles could be used in cooking.

Mount Holyoke College

Preacher Teachers

The French visitor Alexis de Tocqueville was astounded—in a land that had no official religion, that allowed freedom of belief—to find American churches so well attended. Tocqueville arrived during the second Great Awakening. It was a time of religious revival, when almost anyone who felt a call to preach could do so at big, emotional camp meetings in frontier Tennessee and Kentucky. When revivalism went east, it spread in New England and New York and through the South.

During the first Great Awakening, back in the 18th century, ministers such as Jonathan Edwards had preached frightening sermons about everyone's sinfulness. The preachers of the second Great Awakening were molded by the new, democratic beliefs. They said people and society could be improved: salvation was up to individuals.

Lyman Beecher was one of the influential ministers of the time. He helped make people aware of the problems that came with industry and big cities. Reform movements sprang from religious revivalism. In 1826, Lyman Beecher founded the American Society for the Promotion of Temperance (temperance means no drinking of liquor). It was one of the first of the social crusades that included abolition, women's rights, workers' rights, school reform, and something new: concern for the poor and handicapped.

Religious camp meetings like this one, where preachers sometimes drove congregations into a frenzy, were also popular social events. Many people traveled miles to reach them and stayed for days.

hours a day. But she couldn't go to college, because there were no colleges for women.

So Mary Lyon founded one. It was Mount Holyoke College, and it opened in Massachusetts in 1837 with four teachers and 116 students. Lyon insisted that it be a democratic school and that the women do their own cooking and cleaning to keep the tuition price low. Because of that, almost anyone who wanted to attend could afford to do so. Most went on to be teachers.

Emily Dickinson came to Mount Holyoke as a student. Dickinson later became one of America's finest poets, but this was her first time away from home and she was homesick (though Mount Holyoke was only about 10 miles from her home in Amherst). "The school is very large," wrote Emily Dickinson to a friend, "quite a number have left, on account of finding the examinations more difficult than they anticipated, yet there are nearly 300 now....Miss Lyon is raising her standard of scholarship a good deal." Emily Dickinson didn't stay at Mount Holyoke. She had poetry on her mind, but many other women stayed and passed those tough examinations. Mary Lyon intended that her college be as

I'm Nobody

I'm Nobody! Who are you?
Are you—Nobody—Too?
Then there's a pair of us!
Don't tell! they'd advertise—
you know!

How dreary—to be—
Somebody!
How public—like a Frog—
To tell one's name—the live-
long June—
To an admiring Bog!

—Emily Dickinson

Some people believe that Emma Willard opened the first college for women in 1821 when she founded the Troy Female Seminary. Others say the seminary was a girl's finishing school, not a college.

demanding as any college for men, and it was.

In Macon, Georgia, Wesleyan College was actually the first chartered woman's college when it opened its doors in 1836.

American women—black and white—wanted to go to college. So did black men. In Ohio, small Oberlin College accepted women and blacks. A few men's colleges, like Bowdoin, Harvard, and Dartmouth, enrolled black male students. Still, in the 19th century, there were many people who really believed that women and blacks weren't meant to learn.

Special Teachers

Horace Mann

Horace Mann was born in Massachusetts, and, though Massachusetts had free public schools, he never went to school for more than 10 weeks a year. That was because he was so poor that he had to work. When he did go to school he didn't like it much, because the teacher beat him with a stick. That wasn't unusual in the 19th century. Teachers were expected to whip their students if they made noise, or wiggled in their seats, or didn't pay attention. Horace Mann thought that was wrong. He didn't want children to get whipped. He wanted to improve schools—especially public schools. He thought our democracy depended on good, free, public schools.

But no one would listen to Mann's ideas; he couldn't express himself well. He needed an education. So he decided to study on his own. He was 20, and for six months he studied hard, very hard. Then he was accepted at Brown University and was graduated from there with high honors. He became a lawyer and a state senator. But when he became secretary of the Massachusetts Board of Education, he was on his way to making history.

In 11 years as state secretary of education, he doubled the money Massachusetts spent on schools, organized the first teacher-training schools, doubled teachers' salaries, improved the curriculum, and made sure that every boy and girl in Massachusetts went to school for at least six months every year. Other states turned to Horace Mann for advice, and he gave it. He has been called the "Father of American Education." Some people say that free, compulsory public education is the greatest of all American inventions. Do you agree?

Mann wasn't the only remarkable 19th-century educator. There were John Swett in California, Catharine Beecher in Ohio, Henry Barnard in Connecticut, Calvin Wiley in North Carolina, and George Atkinson in Oregon—to mention just a few people who might be worth looking up and writing about.

William H. McGuffey was the most influential of all the educators of his century. He was born in Pennsylvania and became a professor in Ohio and in Virginia. It was the reading books that he compiled that were influential. He called them *Eclectic Readers*—everyone else called them *McGuffey Readers* (what is *eclectic*?). They were phenomenally successful—122 million copies are said to have been sold. For about 50 years, most schoolchildren read them. They gave the nation common stories and a common curriculum.

An early McGuffey Reader *illustration*

McGuffey thought the elementary grades were the most important school years. "The child," he wrote, "needs more help than the boy, and the boy more than the man; hence the primary school requires the best...." He tried to provide the best by including selections in his readers from very good authors. In those days, some schools didn't have libraries; many had poor libraries. McGuffey's books were very important. A few schools still use them today.

21 Do Girls Have Brains?

In 1826 Boston started a girls' high school. It was closed two years later for being "an alarming success."

Women's brains are smaller than men's. Girls can't learn as much as boys. That's what some 19th-century experts actually said. Black men's brains are even smaller, they said, and as for black women—well, it would be silly to attempt to teach them anything.

Who were the experts who were saying those kinds of things? Why, male experts, of course. Some 19th-century white men really thought that nonsense was the truth. A few of those experts were professors, and they were so convincing that some women even believed them.

What about Phyllis Wheatley, the black woman who had learned Latin and written poems that were read in faraway England? Or Lucy Prince, another poet, born a slave, who at age 67 persuaded a governor's council in Vermont that the town of Guilford needed to protect her family after tearing her fence down? Chief Justice Salmon P. Chase said that Prince "made a better argument than any he had heard from a Vermont lawyer."

"Who is Phyllis Wheatley? Who is Lucy Prince?" men asked. History books were written by men. They told

In 1820 Ruth Titus stitched a sampler glorifying the approved role for women: "To give domestic life its sweetest charm." Thirty years before, Judith Sargent Murray wrote: "From what does this superiority [of men] proceed?...the one is taught to aspire, the other is early confined and limited." (Which was which?)

Around 1820 a student at a seminary for young ladies made these sketches of her classmates working and playing. Geography was called "the use of the globes."

Women as Equals

The Grimké sisters ran their own school, where they tried to put their ideas about fairness and equality into practice. Sarah Grimké wrote Letters on the Equality of the Sexes and the Condition of Women. *Here is some of it:*

I spent the early part of my life among the butterflies of the fashion world. And, of this class of women, I must say, their education is miserably deficient. Their chief business is to attract the notice and win the attention of men. They seldom think men will be allured by intellectual achievements, because they find that where mental superiority exists, a woman is generally shunned.

Woman, said Sarah Grimké (right; Angelina Grimké is at left), has "surrendered her dearest rights."

The general opinion that women are inferior to men also has a tremendous effect on women who work....A male teacher can command a higher price than a woman even when he teaches the same subjects and is not in any respect superior to the woman. The same is true of every occupation in which both sexes engage.

Men may reject what I say because it wounds their pride but I believe they will find that woman as their equal is unquestionably more valuable than woman as their inferior both as a moral and intellectual being.

about men. After the Civil War few people remembered Phyllis Wheatley. Few had heard of Lucy Prince.

So, with rare exceptions, colleges would not take women, and there was no way for most women to become doctors, or lawyers, or skilled workers.

If a woman did get a job, she knew that her pay would be about half that of a man doing the same work. Her salary would belong to her husband. Women couldn't vote, so they couldn't change the laws.

A woman married to a bad man was not much better off than a slave. A husband could whip his wife and the law was on his side. A married woman owned nothing. Everything she had—even money from her parents—belonged to her husband. The law made it almost impossible to get a divorce, and when a woman did manage it, many people considered such a thing shameful—although those who divorced were usually desperate. If a wife ran away from her husband, the law said he had rights to the children—even if he was an alcoholic who beat them. If he killed someone in the family, then the law would step in. It was a bit late.

Most people—then and now—are decent. Most 19th-century husbands weren't bad. Most loved their wives and children and treated them well. But even good men were horrified at the idea of a woman standing up and speaking her mind.

"It's a man's world," they said, and in the 19th century it was.

Some women, however, didn't believe their brains were small. They knew they were as smart as any man; they were determined to use their brains.

Sarah and Angelina Grimké had something to say and they said it to audiences of men and women. The sisters set out on a lecture tour in 1837 to tell what they had seen of slavery.

"What kind of women would speak in front of men?" people asked. "Monster women!" That is what some people said of the Grimké sisters. Ministers said it was wrong for a woman to speak when men were there to hear. Surely the Grimké sisters knew of Adam and Eve, and that Eve should have kept quiet. Were they told of Anne Hutchinson and of what happened to her in the Massachusetts Bay Colony?

Those who came to hear the Grimkés saw two women dressed in simple gray Quaker garb. Southern women they were, women who

By the 1850s, some schools, like this one in Boston, were teaching girls serious subjects. As the population grew, women were needed as teachers—and so had to be well educated.

Calisthenics was thought "a much more perfect system" than dancing as a way for modern young ladies to exercise.

Anne Hutchinson was a Puritan woman with strong, questioning ideas. She spoke out and people listened. Some men listened. As you might guess, that disturbed the leaders of the community.

could no longer live in their slave-owning families and who had come north and told why. They told of black children being sold away from their families, they told of whippings with lashes, they told of other horrors. And their audiences were astonished, for they did not know of these things. Angelina went before the Massachusetts legislature and presented tens of thousands of antislavery petitions that had been collected by women. She was the first American woman to address a legislative body.

The ministers were angry. It was not because of what the Grimké sisters said, but because women were saying it. A letter was read in every Congregational church in Massachusetts. The letter said that woman

> *depends on the weakness which God has given her for her protection....but when she assumes the place and tone of man as a public reformer, our care and protection of her seem unnecessary. If the vine...thinks to assume the independence of...the elm...[the vine] ...will fall in shame and dishonor into the dust.*

Now who was the vine and who was the elm? (You may have to read those sentences a few times to figure that out.)

Some women didn't want to be vines. They thought they could stand without support. Half a century earlier, Abigail Adams had been such a woman. Like most women of her time, Adams never went to school. (Remember, she grew up in the 18th century.) She learned to read and write at home. Because she was smart and industrious and kept reading, she soon knew more than most of the school-taught boys.

Just a Joke

Elizabeth Blackwell applied to 29 medical schools before she was accepted, in 1847, at Geneva College (now Hobart College), a small school in upstate New York. Later, Elizabeth found out that the administration had asked the college students to decide whether to admit a woman. The students thought the whole thing was a joke, and so joined in the joke by accepting her. Here are some words from Blackwell's medical school journal:

November 15.—Today, a second operation at which I was not allowed to be present. This annoys me. I was quite saddened and discouraged by Dr. Webster requesting me to be absent from his demonstrations. ...I wrote to him hoping to change things....

Elizabeth Blackwell in old age.

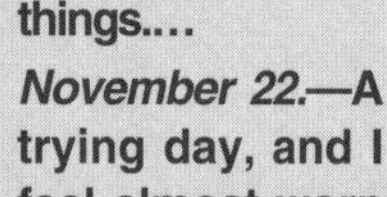

November 22.—A trying day, and I feel almost worn out, though it was encouraging too, and in some measure a triumph; but 'tis a terrible ordeal! That dissection was just as much as I could bear. Some of the students blushed, some were hysterical, not one could keep in a smile, and some who I am sure would not hurt my feelings for the world if it depended on them, held down their faces and shook....I sat in grave indifference, though the effort made my heart palpitate most painfully....

November 24.—Today the Doctor read my note to the class. In this note I told him that I was there as a student with an earnest purpose, and as a student simply I should be regarded; that the study of anatomy was a most serious one, exciting profound reverence, and the suggestion to absent myself from any lectures seemed to me a grave mistake....I listened joyfully to the very hearty approbation with which it was received by the class, and then entered the amphitheatre and quietly resumed my place. The Doctor told me he felt quite relieved.

When Abigail's husband John was making laws for a new nation, she wrote to him:

In the new code of laws...I desire you would remember the ladies...[We] will not hold ourselves bound by any laws in which we have no voice or representation.

There were others who believed as she did.

Some, like Elizabeth Blackwell, who was the first woman to go to a medical school in the United States, proved they had brains as large as anyone's.

But even at Oberlin College, where they were daring enough to educate men and women together, they didn't let women speak out. Only the male students at Oberlin College got to read their graduation essays aloud. If you were a female student you had to let a male professor read your paper for you. Lucy Stone objected. When the other students heard of her protest they agreed with her. But it didn't matter. In 1847, on graduation day, Lucy Stone was not allowed to read her own essay. Some women believed that they needed to change those words *All men are created equal* to *All men and women are created equal.*

Together but Equal

When Henry Blackwell heard Lucy Stone speak before the Massachusetts legislature, he fell in love. Most men were horrified at the idea of women speaking out and wanting to vote. But not Henry. He was a reformer, and he came from an unusual family. Two of his sisters were doctors: Elizabeth (see facing page) was the first woman to graduate from an American medical school. Another sister was a newspaper correspondent, another was an artist, and Henry's brother Sam's wife was Antoinette Brown, who was a woman minister (and that was mighty unusual). So it isn't surprising that a woman who was independent and was fighting for her rights would appeal to Henry Blackwell. But there was a problem: Lucy Stone was determined to devote her whole life and attention to women's rights, and she believed she couldn't do it as a married woman—not in 19th-century America. She said no to Henry. But he was in love, and he wouldn't give up. So, for two years, he kept asking, until, finally, Lucy said yes. According to the laws of the time, a woman had to give up her legal identity and the right to all of her own property when she married. Lucy wouldn't do that. She said she intended to be an equal partner in this marriage. She refused to change her name. That was fine with Henry, so they wrote their own marriage vows. Here is a part of them:

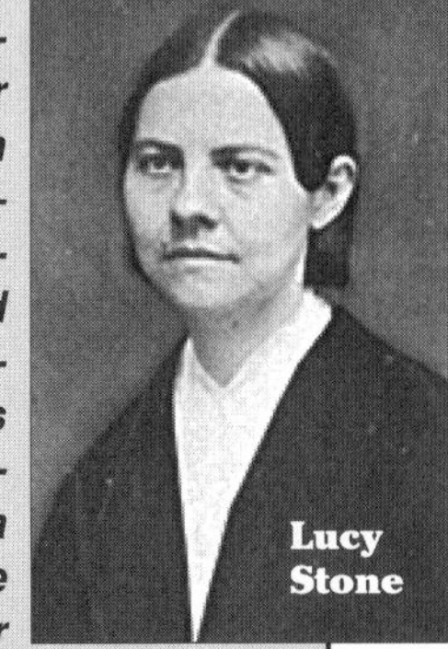
Lucy Stone

While acknowledging our mutual affection by publicly assuming the relationship of husband and wife, yet in justice to ourselves and a great principle, we deem it a duty to declare that this act on our part implies no sanction of, nor promise of voluntary obedience to such of the present laws of marriage, as refuse to recognize the wife as an independent rational being....We believe...that marriage should be an equal and permanent partnership, and so recognized by law. [*Sanction* means approval; *rational* means thinking.]

How did the marriage work out? Very well. It was a happy marriage of two independent rational beings.

Women were taught to draw, but their teachers were usually men.

22 Seneca Falls and the Rights of Women

The childhood of woman must be free and untrammeled. The girl must be allowed to romp and play, climb, skate, and swim; her clothing must be more like that of the boy—strong, loose-fitting garments, thick boots, etc., that she may... enter freely into all kinds of sports.

—ELIZABETH CADY STANTON, 1851

The idea of women's rights—and women wearing pants—was ridiculed in cartoons like this one.

Elizabeth Cady read the nation's great Declaration, and it bothered her. *All men are created equal,* it said. But what about all women? Elizabeth's father, Daniel Cady, was a judge; she spent hours in his office listening and learning law. "If only you had been born a boy," he told her. "You could have been a lawyer."

Elizabeth didn't want to be a boy; she was happy being a girl. But she wanted to use her mind—and she did. She decided to learn everything the boys were learning. She asked a scholarly neighbor to teach her Greek. He did, and she learned well. At school she was a top student. But when graduation came and the boys all went off to college, Elizabeth couldn't go with them, because she was a girl. She was sent to Emma Willard's seminary, a finishing school for girls. She couldn't be a lawyer, because women weren't allowed to practice law.

Elizabeth was determined to do something with her life; it helped that she married a fine man. He was Henry B. Stanton, a leader of the antislavery movement. When they married, the minister expected her to promise to "obey" her husband. That was a

Lucretia Mott said, "Let woman...receive encouragement for the proper cultivation of all her powers." What did she mean by that?

customary part of the marriage ceremony. Elizabeth Cady wouldn't use that word. "I obstinately refused to obey one with whom I supposed I was entering into an equal relation," she wrote. The minister was unhappy, but Henry understood. He was marrying a woman who wouldn't do things that didn't make sense, even if everyone else did them.

Elizabeth Cady Stanton was soon using her intelligence to help women. She and Lucretia Mott and other friends decided to organize a women's rights convention—the first in the nation. It took eight years of planning, but, finally, in July 1848, some 300 people—men and women—met in the Methodist Church in Henry Stanton's hometown: Seneca Falls, New York. They wrote a declaration; it is known as the Seneca Falls Declaration, and it says, *We hold these truths to be self-evident: that all men and women are created equal.*

Cautious, careful people, always casting about to preserve their reputation and social standing, never can bring about reform.

—SUSAN BROWNELL ANTHONY, 1860

Then, using Jefferson's Declaration of Independence as a guide, the women and men at Seneca Falls went on from there, telling of all the ways they felt women were being wronged. Thomas Jefferson's words accused King George III of tyranny; Elizabeth Stanton's accused "man." One hundred people signed the Seneca Falls Declaration; 32 of them were men.

Did anyone pay attention to that Declaration? Yes, indeed. This is what Elizabeth Cady Stanton wrote in her autobiography:

> *No words could express our astonishment on finding, a few days afterward, that what seemed to us so timely, so rational, and so sacred, should be a subject for sarcasm and ridicule to the entire press of the nation.*

She continued:

> *All the journals from Maine to Texas seemed to strive with each other to see which could make our movement appear the most ridiculous. The anti-slavery papers stood by us...and so did Frederick Douglass [the famous abolitionist who had been born a slave]...but so pronounced was the popular voice against us, in the parlor, press, and pulpit, that most of the ladies who had attended the convention and signed the declaration, one by one, withdrew their names and influence and joined our persecutors.*

Elizabeth Cady Stanton may have been discouraged, but that didn't stop her. She had started

Insane Concerns

Dorothea Dix was a lonely child. Her mother was dead, and her father was a traveling preacher. Dorothea was sent from relative to relative to live. Perhaps that was why she had sympathy for those who were troubled—especially for the mentally ill. People with mental or emotional problems were being put in privately run, unregulated mental asylums [uh-SY-lums], and they were usually horrible places. In 1843, Dix wrote a *Memorial to the Legislature of Massachusetts* telling of "insane persons confined within this Commonwealth, in cages, closets, cellars, stalls, pens! Chained, naked, beaten with rods, and lashed into obedience." Hardly anyone then concerned themselves with prisons, or poor houses, or mental asylums. For more than 30 years, Dix traveled about the country, and to other nations, too, visiting asylums and prisons, and writing and speaking about them.

Dorothea Dix

Reforming Zeal

Julia Ward Howe and her husband, Samuel Gridley Howe, were an amazing couple, even if they didn't always get along. Both were reformers, intent on improving the world.

Julia is most remembered for a poem she wrote, the "Battle Hymn of the Republic." She was a writer, and a founder and president of the New England Woman Suffrage Association (1868).

Sam was a doctor, but he is most remembered as founder of the Perkins Institute, the first school in the United States for the blind. When Sam taught Laura Dewey Bridgman (below) to read and write, he astonished most educators. Laura was both deaf and blind. No one knew how to teach her. But Samuel Howe put raised letters on spoons and cups and other familiar objects. Laura learned to read the names with her fingers.

Both Howes were abolitionists, and they edited an abolition paper together. Sam worked to improve public-school education and to better conditions for the mentally ill. Maybe it was because the Howes were both strong-willed that they often argued. Samuel Howe didn't think a woman should have a public career. Julia did. Julia wrote, "I have never known my husband to approve of any act of mine which I myself valued. Books—poems—essays—everything has been contemptible in his eyes because not his way of doing things." Julia Ward Howe was the first woman to be elected to the American Academy of Arts and Letters, but that was after Sam was dead.

something that would grow and grow. "We had set the ball in motion, and now, in quick succession, conventions were held in Ohio, Indiana, Massachusetts, Pennsylvania, and in the City of New York."

The men and women who met in those conventions, and all those who worked for women's rights, were called reformers. Most reformers, in the mid-19th century, were churchgoers who were fighting for all these things: abolition (to end slavery), temperance (to ban the drinking of alcoholic beverages), and women's and children's rights. (Later, those movements all split apart.)

Amelia Bloomer said, "Men call us angels... but at the same time they are...subjecting us to virtual slavery."

Amelia Bloomer, who was editor of a temperance newspaper, was one of the reformers. She said women should get out of the long garments that made them trip—or even faint when they were laced too tight at the waist. Amelia Bloomer wore long pantaloons (wide trousers) under her short dresses, and tried to talk other women into that fashion. It allowed for freedom of movement, she said. But that was going too far for most 19th-century Americans. People threw stones at the "bloomer girls" and made fun of them. It was a long time before women felt comfortable wearing pants.

One day Amelia Bloomer introduced Elizabeth Cady Stanton to Susan B. Anthony. That was in 1851. It was a momentous meeting. Stanton and Anthony formed a team—like Lewis and Clark—and they left a big imprint on American history.

Elizabeth Cady Stanton (left) and Susan B. Anthony. Stanton died in 1902; Anthony in 1906. It was not until 1920 that women got the vote.

They were an uncommon

It became a popular sport for some men to disrupt women's rights conventions such as this one, held in 1857—and for cartoonists to make fun of the women participants.

REPORT

OF THE

WOMAN'S-RIGHTS CONVENTION,

HELD AT

Seneca Falls, N. Y.,

JULY 19TH & 20TH, 1848.

ROCHESTER:
PRINTED BY JOHN DICK,
AT THE NORTH STAR OFFICE.
1848.

The original title page of the Seneca Falls Declaration.

pair, with talents that complemented each other. Elizabeth Cady Stanton was chubby and merry, and the mother of a family that kept growing and growing. (She had seven children altogether.) Susan Anthony was slim and reserved, and she never married. When she spoke she made sense.

In 1853, Anthony attended a teachers' conference in Rochester, New York. She had been a teacher for 15 years, and she asked for permission to speak at the meeting. No woman had done that before. It took the men in charge a half hour to decide. Then they agreed to hear her. This is part of what she said:

> *Do you not see that so long as society says a woman has not brains enough to be a doctor, lawyer or minister, but has plenty to be a teacher, every man of you who chooses to teach...admits...that he has no more brains than a woman?*

Her audience was shocked (and that was what she intended).

Susan B. Anthony was a determined woman, a superb organizer, and the kind of person who never gives up. Together with Elizabeth Cady Stanton, she led the movement that eventually brought the vote

For many people, the notion of women "wearing the pants" in marriage (above: *Man as He Expects to Be*) or in society (below, the artist's view of *The Triumphs of Women's Rights*) seemed like the collapse of civilization.

Climb Every Mountain

Julia Archibald Holmes and her husband, James Holmes, heard that there was gold near Pikes Peak in Colorado. Remember Pikes Peak? It was the mountain Zebulon Pike tried to climb and couldn't. And that some of Stephen Long's men tried to climb and did.

The Holmeses joined a wagon train of several dozen gold seekers heading west from Kansas. Two were women. Julia wore what she called an "American costume"—short dress, bloomers, moccasins, and a hat. "It gave me freedom to roam at pleasure in search of flowers and other curiosities, while the cattle continued their slow and measured pace." The other woman disapproved.

That didn't bother Julia. She was determined to keep up with the men on the trip. So she trained by walking—farther and farther each day—until she could easily cover 10 miles in the hottest weather. "Believing as I do, in the right of woman to equal privileges with man, I think that...we should...share the hardships which commonly fall to the lot of man." So she asked to share guard duty with the men. She was turned down.

But when it came to climbing the peak, she would not be turned aside. On August 1, 1858, Julia, James, and two men set out. They

to women (although that didn't happen until the 20th century, after both of them were dead).

In the 19th century, many people believed a man represented his whole family when he voted. If women were given the vote, they might disagree with their husbands and vote differently. Would that break up the family? Men and women worried about that. Even Stanton's friend Lucretia Mott wasn't sure that women should vote. She just wanted them to have equal rights. But could they have equal carried backpacks that held 19 pounds of bread, six quilts, a pound of pork, a pound of sugar, some coffee, plates, forks, a canteen, cups, writing materials, and a book of essays by Ralph Waldo Emerson. Julia wore her American costume. It was a tough climb. The first night they camped by a stream and built a fire.

> *There is music from the foaming stream, sounds from a dozen little cascades near and far blend together—a thundering sound, a rushing sound, a rippling sound, and tinkling sounds...the burning pine crackles and snaps, showering sparks, cinders and even coals around and all over the sheet I am writing on.*

The next day they were up early. This time they climbed into a snow squall. Julia wrote of their next camp:

> *We have given this name [Snowdell] to a little nook we are making our home in for a few days....On the cold moss overhung by two huge rocks, forming a right angle, we have made a nest of spruce twigs. Some smaller rocks form, with the larger ones just mentioned, a trough about three feet wide, and ten feet long. At the outlet of this narrow space we have built a chimney. When we lie down the fire is burning but a yard from our feet....The beauty of this great picture is beyond my powers of description. Down at the base of the mountain the corral of fifteen wagons, and as many tents, form a white speck.*

On the morning of August 5 they headed for the summit taking only "writing materials and Emerson." When they reached the goal, Emily read a poem and wrote letters to friends, "using a broad flat rock for a writing desk." They were 14,110 feet above the level of the ocean. It was soon snowing hard, but Julia felt triumphant.

> *I have accomplished the task which I marked out for myself....Nearly everyone tried to discourage me from attempting it, but I believed that I should succeed....In all probability I am the first woman who has ever stood upon the summit of this mountain and gazed upon this wondrous scene.*

The Holmeses moved to Santa Fe. James became secretary of the Territory of New Mexico; Julia became a correspondent for the *New York Tribune*. Many years after that she was chief of the Division of Spanish Correspondence for the Bueau of Education in Washington, D.C, and was active in the women's suffrage movement.

Unlike Julia Holmes, these ladies made it to the top wearing skirts.

rights if they couldn't vote?

What about women who weren't married? What rights did they have? How could they take part in a democracy? And what about women who did disagree with their husbands? Anthony and Stanton knew that without a vote women were like helpless vines clinging to men.

23 A Woman Named *Truth*

The chairwoman of the Ohio convention wrote: "I have never...seen anything like the magical influence that...turned the sneers and jeers of an excited crowd into notes of respect and admiration."

The man at the women's rights convention in Ohio, in 1851, thought he was telling the truth when he said that women were by nature weak and inferior to men. But a lean, stately woman who heard him didn't agree. She was almost six feet tall and she wore a gray dress with a white shawl; a white turban was wrapped around her head; when she walked it was with the dignity of a queen.

The tall woman was the only black person in the church where the convention was held—and all eyes were upon her. Finally she could bear no more. Her name was Sojourner (SO-jer-ner) Truth, and she stood up and spoke in a voice like rolling thunder. The hall was hushed, and that voice rang out and asked:

> *Ain't I a woman? Look at me. Look at my arm. [And she showed powerful muscles.] I have ploughed, and planted, and gathered into barns, and no man could head me! And ain't I a woman? I could work as much and eat as much as a man—when I could get it—and bear the lash as well! And ain't I a woman? I have borne thirteen children, and seen them most all sold off to slavery, and when I cried out with my mother's grief, none but Jesus heard me! And ain't I a woman?*

She said more, and said it powerfully. A woman who heard her wrote:

> *Amid roars of applause, she returned to her corner, leaving more than one of us with streaming eyes, and hearts beating with gratitude. She had taken us up in her strong arms and carried us safely...turning the whole tide in our favor.*

The abolitionists and those who were working for women's rights often had close ties. Feminist Susan B. Anthony was a paid agent for the American Anti-Slavery Society, and the abolitionist Frederick Douglass spoke at women's rights meetings.

Sojourner Truth had been named Isabella when she was born a slave in New York State. She was treated harshly, as slaves often were. In 1826, the year before New York freed its slaves, she ran away. She was a young mother and she planned to buy her children. But, before she could, her former master sold one of her children—to a buyer in the South. That was against New York law. With the help of a white Quaker family she found a lawyer, went to court, and won the child! That said much for her determination, and also for the fairness of the court. Later, she helped other blacks go to court and fight for their rights.

Isabella's Quaker friends told her, "Before God, all of us are equal." No one had ever said that to her before. She could not read, but they read to her from the Bible. She soon memorized large parts of the Bible. She became deeply religious and had visions of God. She decided to live a godly life and to help others. Isabella chose a new name to celebrate her freedom and her new way of life. It was Sojourner Truth. A *sojourner* is a traveler who stops somewhere for a short time and then continues on. For the next 40 years she traveled and spoke out for truth and justice.

Perhaps it was her dignity, or her sincerity, or that mighty voice, but when Sojourner Truth spoke people listened. Across her chest she wore a banner that said, PROCLAIM LIBERTY THROUGHOUT ALL THE LAND UNTO ALL THE INHABITANTS THEREOF. Those words from the Bible are written on the Liberty Bell in Philadelphia.

One abolitionist, Lydia Maria Child, wrote, in 1833, *An Appeal on Behalf of That Class of Americans Called Africans.* Here are some of her words: *They [the slaves] have stabbed themselves for freedom—jumped into the waves for freedom—starved for freedom—fought like very tigers for freedom! But they have been hung, and burned, and shot—and their tyrants have been their historians!*

Sojourner Truth soon became famous. Harriet Beecher Stowe, a well-known writer (whom you'll read about in Book 6 of *A History of US*), was her friend, and Abraham Lincoln invited her to the White House. She spoke out against injustice, wherever she found it. She worked for women's rights, black rights, prison reform, and temperance. Once, a man tried to make fun of her, saying, "I don't care any more for your talk than I do for the bite of a flea."

Sojourner Truth chuckled as she replied, "Maybe not, but the Lord willing, just like the flea, I'll keep you scratching."

Nineteenth-century women did keep people scratching. They were working in factories, speaking in public, writing for newspapers, and fighting for causes they believed in.

Harriet Beecher Stowe might have looked small and frail, but she was tough enough to start writing her famous book *Uncle Tom's Cabin* shortly after the birth of her seventh child.

24 Life in the Mills

An ironworker brings a load of ore to the furnace. Most Americans still worked on the land, but more and more went to the mills and mines.

No people had ever inhabited a land that offered so much to so many. Most Americans lived on farms, got up early in the morning, and went cheerfully to work: milking cows, feeding chickens, building houses, growing crops, and raising children. Ten children was not unusual for a 19th-century American family. However, big as they were, families were smaller than they had been in the 18th century. They would continue to get smaller. No one was quite sure why. One thing was clear. For most people, life was good in America. The air was pure, the streams full of fish, the woods full of game, the meadows full of wildflowers. Besides, we were a free people, working for ourselves.

In 1820, half of the nation's industrial workers were children under 10 years of age. However, the heavy iron-foundry work took men's muscles.

Well, most of us were. Two groups were not. There

The heat and vibration and noise—especially the noise—of the steel mills were often compared to the torments and fires of hell.

were the slaves, who sometimes worked from sunup to sundown, and only rarely for themselves. Free people were beginning to have guilty feelings about slavery. A few people—abolitionists—were working to end the horrors of black slavery.

There was another group—also enslaved—but not by law or skin color. They were slaves of factory, mine, and mill. Their lives were horrible. They had no fresh air to breathe. They went into the mills as children. Often, they died in their thirties. They were almost without hope. Hardly anyone knew they existed.

Then, in 1861, an article appeared in the most important magazine of the day: the *Atlantic Monthly*. The article, called "Life in the Iron Mills," was unsigned. It created a sensation. Everyone began talking about it—but no one suspected that the writer was a shy, unmarried, 30-year-old woman. Her name was Rebecca Harding.

Harding lived in Wheeling, Virginia. (Soon Wheeling and the surrounding area would break away from Virginia and become West Virginia.) Wheeling was an Ohio River town, on the border between North and South. Some white people there kept slaves and believed in slavery, but most did not. It was a border in other ways, too. The rough-and-tumble frontier was just spitting distance away. Not far off—in the other direction—were the planter societies of Virginia and Maryland.

Thirteen thousand people lived in Wheeling. Some of them were wealthy and prosperous; their gorgeous homes lined the riverfront. Wheeling had industry; it had farming nearby, and it was a transportation hub. The nation's highway, the National Road, went right through the town. From Wheeling you could take a riverboat

Striking Times

"American Ladies Will Not Be Slaves," read the Lynn shoemakers' banner. "Give Us a Fair Compensation and We Labor Cheerfully."

It was a snowy day in March 1860 when the women shoemakers of Lynn, Massachusetts, went on strike. The weather didn't discourage them. They took out their parasols, put on their hoop skirts, marched behind the Lynn city guard and its brass band, and sang. Most of the town turned out to watch and sing with them. Two weeks later the strike was still underway and the women had a chowder party to keep up their spirits (with dancing and kissing at the party). But when one of their bosses said he had persuaded some women to come back to work, the strikers weren't amused. They kicked the boss out of the party and had the band play the "Rogue's [say *roegz*] March." The striking women were serious about their goal: higher wages. Some people (especially employers) didn't like the idea of labor unions or strikes, but strikes were judged legal by Chief Justice Lemuel Shaw of the Massachusetts Supreme Court in 1842. (*Labor unions* are organizations of workers; *strikes* are work stoppages to protest low wages or bad working conditions; a *rogue* is a villain.)

to St. Louis and New Orleans and then on around Florida to the East Coast. When Rebecca Harding was a girl, if you wanted to cross the river you had to take a ferry. But in 1849, a year after she graduated from high school, the world's largest single-span bridge—the Wheeling Suspension Bridge—crossed the Ohio River. It became the town's pride.

Rebecca lived in one of those grand houses on the river. She was valedictorian—top student—of her class. She wanted to go on to college, like her brother, but because she was a girl, she knew it wouldn't happen. So Rebecca turned to her books. As long as she could remember, she had lived in the imaginary world that books create. She loved fairy tales and stories of knights, princesses, goblins, and wizards. She began to write novels and short stories. Then one day she took a new book to her hideaway in a cherry tree. The book was by an American author named Nathaniel Hawthorne. She was amazed by it. The author had written about ordinary people, in the United States, and about a girl like herself. It made her realize that "the commonplace folk and things which I saw every day...belonged to the magic world [of books] as much as knights and pilgrims."

It was the commonplace folk of Wheeling she wrote about in the article that stunned all those who read it. She wrote of people whose lives were so different from those in the big houses they might as well have lived in a different galaxy.

Pittsburgh (top) became rich because western Pennsylvania was the heart of the coal and iron industries in the 19th century. But what was it like for those who lived out their lives under the smoke that "rolls sullenly in slow folds from the great chimneys of the iron foundries"? Below, workers pour ingots in a steel mill.

Wheeling was an iron-mill town. Ironworkers dragged themselves to work that lasted 14 hours a day, six days a week. They went into the mill at age nine or ten, and left when they died.

Rebecca Harding wrote of the ironworkers and their town—it was her town too. She described the smoke that

> *rolls sullenly in slow folds from the great chimneys of the iron foundries, and settles down in black, slimy pools on the muddy streets. Smoke on the wharves, smoke on the dingy boats, on the yellow river,—clinging in a coating of greasy soot to the house-front, the two faded poplars, the faces of the passers-by.... Smoke everywhere! A dirty canary chirps desolately in a cage beside me. Its dream of green fields and sunshine is a very old dream,—almost worn out, I think.*

In a country where the idea of pollution was still unknown, where sunshine and clear air were believed to be nature's gifts to all, Harding's description was startling.

She was just beginning. She went on to tell of the people who worked in the hot furnace of a mill.

> *Masses of men, with dull, besotted faces bent to the ground, sharpened here and there by pain or cunning; skin and muscle and flesh begrimed with smoke and ashes; stooping all night over boiling cauldrons of metal...breathing from infancy to death an air saturated with fog and grease and soot, vileness for soul and body.*

Rebecca Harding told the story of a young ironworker: Hugh Wolfe. Hugh was fictional, but he might have been real. He was talented—he made statues of men and women out of the refuse from the iron mill, and he dreamed of life as an artist. Hugh's poor, homely cousin Deborah wanted to make his dream come true. So she stole money and gave it to Hugh. He got caught and sent to prison, where he died, not yet 20.

People wept when they read the story, and learned—often for the first time—of the wage slaves who tended the scalding pots of liquid metal that became the iron and steel needed to build the railroads and machines the nation was demanding.

Did her story help those workers? Probably not, for no one then knew how to smelt iron, or dig minerals, or make steel without hard, horrible, backbreaking labor. No one wanted to do that kind of labor. No one should have had to do it. The carbon monoxide vapors from the burning coal destroyed people's lungs. The liquid metal spilled, hardened, and flew off as accidental bullets. A visitor to nearby Pittsburgh described that milltown as "Hell with the lid taken off." Someday the problem would be partly solved by machines. But in the 19th century no one knew that. Harding's story made some people of her time aware of the horror of millwork.

Did Rebecca Harding write anything else? Did she ever leave Wheeling? Yes, to both of those questions. She wrote many articles, and some books too. Mostly, she wrote of people with dreams and longings. One of her own dreams came true when she met Nathaniel Hawthorne. Remember, it was Hawthorne's stories that had made her want to be a writer.

A man named L. Clarke Davis read and admired her stories. He asked to meet her and they fell in love, married, and had a son, Richard Harding Davis, who became the most famous war correspondent of his day. But he never wrote of working people and their problems, as his mother had. Few writers did.

The broad Ohio River was the perfect highway for those restless Americans who wanted to go west. You could get on a flatboat in an Appalachian mountain town and float down to the Mississippi—or get off any place along the way. That rich land west of the Appalachians and east of the Mississippi was filling up with farms and families.

Ponies and boys in the coal mines. Pennsylvania was estimated to have 10,000 square miles of coalfields. In 1820 the state's coal output was 365 tons. In 1848 it was almost a million and a half tons.

25 Working Women and Children

A ***phenomenon*** (fuh-NOM-ih-nun) is a fact or event, especially one that you can see for yourself.

A pioneer woman and her tools—apron and broom. Keeping a bare-earth or split-log floor clean and swept needed constant work.

Women in early America didn't work.

Do you believe that? Some people do. They think working women are a new phenomenon. Well, people who believe that don't know their history. Women have always worked in America. They just didn't always work for wages.

How about American men? Have they always worked? Of course. But they didn't work for wages either. In the old days, most men and women worked on farms, for themselves, and didn't see much cash. If they had extra crops, they sold or bartered them for goods they couldn't produce themselves.

Some American men and women were indentured servants or slaves, and of course they didn't work for themselves, and they usually didn't get wages either. A few people, however, were paid cash or wages for their work: craftspeople (blacksmiths, shoemakers, furniture makers, printers, etc.), teachers, and ministers.

In Europe those cash-paying jobs had almost always belonged to men. America was different. From the beginning there was a labor shortage. If a blacksmith wanted to teach his daughter his craft, he could do it in America. Some American women were blacksmiths and shoemakers and coopers (barrelmakers). Some were teachers. The traditional churches would not ordain women as ministers, but on the frontier—where preachers were in short supply—women were often called to preach.

The noise and commotion of the New England spinning machines were hard to get used to—"the buzzing and hissing and whizzing of pulleys and rollers and spindles and flyers."

When factories began to open in America, women filled the factories. There was an unfortunate reason for that. Women were willing to work for lower wages than men. There was another group that worked for still lower wages: children. Nineteenth-century factories were filled with children—boys and girls who almost never got a chance to play.

Do you sometimes complain about school? Well, you might stop complaining after you read what factory life was like for some 19th-century children. This description was written by Herman Melville in 1855. (You'll hear more about Melville in a few chapters.) Melville needed some paper (he was a writer), so he got on his horse and went off to a paper factory. He made the trip down a mountain ravine into a hollow called Devil's Dungeon. It was a cold, cold winter day, and his cheeks were bitten with frost when he arrived.

Children like this boy on a spinning machine had three times as many accidents as adults.

TIME TABLE OF THE LOWELL MILLS,

To take effect on and after Oct. 21st, 1851.

The Standard time being that of the meridian of Lowell, as shown by the regulator clock of JOSEPH RAYNES, 43 Central Street

	From 1st to 10th inclusive.				From 11th to 20th inclusive.				From 21st to last day of month.			
	1stBell	2dBell	3dBell	Eve.Bell	1stBell	2d Bell	3d Bell	Eve.Bell	1stBell	2dBell.	3dBell	Eve.Bell.
January,	5.00	6.00	6.50	*7.30	5.00	6 00	6.50	*7.30	5.00	6.00	6.50	*7.30
February,	4.30	5.30	6.40	*7.30	4.30	5.30	6.25	*7.30	4.30	5.30	6.15	*7.30
March,	5.40	6.00		*7.30	5.20	5.40		*7.30	5.05	5.25		6.35
April,	4.45	5.05		6.45	4.30	4.50		6.55	4.30	4.50		7.00
May,	4 30	4.50		7·00	4.30	4.50		7.00	4.30	4.50		7 00
June,	"	"		"	"	"		"	"	"		"
July,	"	"		"	"	"		"	"	"		"
August,	"	"		"	"	"		"	"	"		"
September,	4.40	5.00		6.45	4.50	5.10		6.30	5.00	5.20		*7.30
October,	5.10	5.30		*7.30	5.20	5.40		*7.30	5.35	5.55		*7.30
November,	4.30	5.30	6.10	*7.30	4.30	5.30	6.20	*7.30	5.00	6.00	6.35	*7.30
December,	5.00	6.00	6.45	*7.30	5.00	6.00	6.50	*7.30	5.00	6·00	6.50	*7.30

* Excepting on Saturdays from Sept. 21st to March 20th inclusive, when it is rung at 20 minutes after sunset.

YARD GATES,

Will be opened at ringing of last morning bell, of meal bells, and of evening bells; and kept open Ten minutes.

MILL GATES.

Commence hoisting Mill Gates, Two minutes before commencing work.

WORK COMMENCES,

At Ten minutes after last morning bell, and at Ten minutes after bell which "rings in" from Meals.

BREAKFAST BELLS.

During March "Ring out"........at....7.30 a. m.........."Ring in" at 8:05 a. m.
April 1st to Sept. 20th inclusive.....at....7 00 " " .... " " at 7.35 " "
Sept. 21st to Oct. 31st inclusive.....at....7.30 " " .... " " at 8.05 " "
Remainder of year work commences after Breakfast.

DINNER BELLS.

"Ring out"...... ....12.30 p. m.........."Ring in".... 1.05 p. m.

In all cases, the *first* stroke of the bell is considered as marking the time.

Lowell mill girls worked 6 days a week, 12 hours a day. Bells rang them to work, meals, and bed, as this table shows.

Dear Father, I am well which is one comfort. My life and health are spared while others are cut off. Last Thursday one girl fell down and broke her neck which caused instant death. She was going in or coming out of the mill and slipped down it being very icy. The same day a man was killed by the [railroad] cars. Another had nearly all of his ribs broken. Another was nearly killed by falling down and having a bale of cotton fall on him. Last Tuesday we were paid. In all I had six dollars and sixty cents paid $4.68 for board [rent and food]. With the rest I got me a pair of rubbers and a pair of .50 cts. shoes....At 5 o'clock in the morning the bell rings for the folks to get up and get breakfast. At half past six it rings for the girls to get up and at seven they are called into the mill. At half past 12 we have dinner are called back again at one and stay till half past seven. I get along very well with my work. I can doff [change the bobbins] as fast as any girl in our room. I think I shall have frames before long. The usual time for learning is six months but I think I shall have frames before I have been in three as I get along so fast. I think that the factory is the best place for me and if any girl wants employment I advise them to come to Lowell.

I Am Well Which Is One Comfort

Mary Paul grew up in northern Vermont and went to work in the Lowell, Massachusetts, textile mills while still in her teens. Mary wrote often to her father to tell of her experiences and get news from home:

Children in textile mills had nimble fingers and could be paid much less than adults.

This is what Melville found:

> *Not far from the bottom of the Dungeon stands a large whitewashed building…like some great white sepulchre, against the sullen background of mountainside firs.…The building is a paper-mill.…immediately I found myself standing in a spacious place, intolerably lighted by long rows of windows, focusing inward the snowy scene without.*
>
> *At rows of blank-looking counters sat rows of blank-looking girls, with blank, white folders in their blank hands, all blankly folding blank paper.…I looked upon the first girl's brow, and saw it was young and fair; I looked upon the second girl's brow, and saw it was ruled and wrinkled. Then, as I still looked, the two—for some small variety to the monotony—changed places; and where had stood the young, fair brow, now stood the ruled and wrinkled one.…Not a syllable was breathed. Nothing was heard but the low, steady, overruling hum of the iron animals. The human voice was banished from the spot. Machinery—that vaunted slave of humanity—here stood*

These women are making matchboxes on an early kind of assembly line.

A ***sepulchre*** is a burial vault or tomb. ***Sullen*** is resentfully gloomy. ***Intolerably*** means unbearably. To ***banish*** means to drive away or exile. To ***vaunt*** means to boast about something. ***Monotony*** means boring, unchanging sameness.

Even in bad factories children were probably better off than in private workshops like this one, doing jobs such as sorting rags or stripping tobacco. If a child fell asleep she could be beaten or have cold water thrown in her face.

menially served by human beings...as the slaves serve the Sultan. The girls did not so much seem accessory wheels to the general machinery as mere cogs to the wheels.

Those girls didn't have a choice. They had to work. Slave children didn't have a choice either. "I was only seven years old when I was sent away to take care of a baby," said Harriet Tubman who, when she grew up, became famous for helping slaves run away to freedom.

One mornin' after breakfast I stood by the table waiting [to clear the table]; just by me was a bowl of lumps of white sugar....I never had anything good; no sweet, no sugar, and that sugar, right by me, did look so nice, and my Missus's back was turned to me so I just put my fingers in the sugar bowl to take one lump, and maybe she heard me for she turned and saw me. The next minute she had the rawhide down [to whip Harriet]...I ran and I ran and I ran.

Menially means acting like a servant. An ***accessory*** is a helpful but not really important or essential addition. What is a ***sultan***? What is a ***cog*** in a wheel?

A wood sawyer and his son eat lunch. Children in the country began working for their parents as soon as they could pick vegetables or carry a bucket.

26 American Writers

The Swedish woman who made this sketch of Ralph Waldo Emerson found him "too cold and hypercritical."

New England was different. New England had always been different. To a visiting New Yorker, everyone in Boston seemed alike. "Why, it is as if they are all of the same family." And—except for the newly arrived Irish immigrants and the African Americans—it was true. In New England, many were descendants of 17th-century Puritans.

That old-time religion was now gone. The sermons of the new churches—Congregationalists, Unitarians, Presbyterians—were less stern and more forgiving. But the Puritan heritage was there. It taught a love of learning, and it taught honesty and a sense of duty. New England children went to school, worked hard, and learned. In Boston and the nearby towns many children learned Latin, and often Greek and Hebrew too. So on Sunday, when the minister told a joke in Latin, everyone laughed.

Boys and girls who went to school in Boston read many of the great Greek and Latin poets; they knew Homer (who wrote in Greek), they knew Virgil (who wrote in Latin), and they knew the English poets. Then some of them began to realize that something was missing. Where were the great American poets?

It was at Harvard—where the brightest of the New England boys went to school—that they started worrying about those missing American writers. Their teachers got them worrying, especially a teacher named George Ticknor. Ticknor had been everywhere, or so it seemed.

He had learned German in Germany, French in France, Spanish in Spain, and Italian in Italy. While he was in all those places he had

James Fenimore Cooper slammed shut the English novel he was reading (Cooper had a temper). "I could write a better book than that!" he told his wife. "Why don't you?" Susan Cooper said. And so began the career of the most popular novelist of his day. Cooper, whose father founded the frontier settlement of Cooperstown, New York (what is Cooperstown famous for today?), was a country gentleman. He soon became a hardworking writer. His tales of frontier heroes and his sea adventures were very successful (though they may seem slow-paced to modern readers). You may have seen the movie based on Cooper's novel *Hawkeye, or The Last of the Mohicans.*

Thomas Jefferson tried to lure Professor Ticknor to the university he founded in Virginia. Ticknor wouldn't budge from Harvard.

Parkman

spoken to the greatest European thinkers and writers. He was astounded at how much they admired America. A famous English poet, Robert Southey, had even written a poem about King Philip's War. No American poet had written about King Philip.

It was time for Americans to write their own stories, said Ticknor. Other Harvard professors said the same thing. They said that to their students who included Ralph Waldo Emerson, Richard Henry Dana, Henry David Thoreau, and Francis Parkman. Those students did just what their teachers expected—they became great American writers.

King Philip's War—between New England settlers and the Wampanoag, Nipmuck, and Narraganset tribes—was fought in 1675 and 1676.

Ralph Waldo Emerson lived in Concord. Do you remember Concord? Think hard. It is the town, 30 miles from Boston, where the first shot of the Revolutionary War was fired. Emerson called it "the shot heard round the world."

In 1850, Concord was a center of American culture. Emerson was known as the "Sage of Concord." (A sage is a wise man.) Henry Thoreau was his neighbor. Nathaniel Hawthorne lived nearby. Hawthorne was a descendant of John Hathorne, who was a judge in the Salem witch trials, which may help explain Nathaniel's haunting stories. Louisa May Alcott was another neighbor. She, of course, couldn't go to Harvard. Women weren't admitted. She did all right on her own. She had to. Her father, Bronson, who was a teacher and a social reformer, was always doing good things for others, but he never worried about earning a living. His large family was often without money. Louisa Alcott realized she would have to be a breadwinner. When she was 16 she published her first book: a collection of fairy tales. When she wrote *Little Women* she solved the family's financial problems. *Little Women* is a novel whose characters are based on Louisa's own family. It is a very good story, both funny and sad. Alcott became one of America's most popular authors.

Hawthorne

Alcott

As the blood of all nations is mingling with our own, so will their thoughts and feelings finally mingle in our literature. We shall draw from the Germans, tenderness; from the Spaniards, passion; from the French, vivacity; to mingle more and more with our English solid sense.

—HENRY WADSWORTH LONGFELLOW, *KAVANAGH*, 1849.

Longfellow

Henry Wadsworth Longfellow (a poet who took Professor Ticknor's place when he retired from Harvard in 1835) often came to Concord to visit. Longfellow could read all the languages that Ticknor read, and Finnish and Swedish too. Most Americans knew Longfellow's poems. Children read them in school, and (in those days before television) grownups read them out loud to each other. They were

translated and admired around the world. Longfellow wrote American poems: about an Indian named Hiawatha, about Paul Revere, about a village blacksmith, about the Acadians who were forced by the British to leave their homes.

It was Emerson who drew people to Concord. He was like a team captain who inspires others. Emerson was tall and skinny with blue eyes and a long nose. He liked to talk and he had interesting things to say. People were attracted to the sage of Concord.

Emerson loved nature; he thought if people learned to understand nature they would understand themselves and many of the secrets of the world. He believed everything has its own place in nature. A fox doesn't try to be a rabbit. The same should be true of people. Find yourself, said Emerson. Do what you can do best. He wrote about his ideas in essays and poems. Here is one of his poems, called "Fable."

The mountain and the squirrel
Had a quarrel,
And the former called the latter "Little Prig;"
Bun replied,
"You are doubtless very big;
But all sorts of weather
Must be taken in together,
To make up a year
And a sphere.
And I think it no disgrace
To occupy my place.
If I'm not so large as you,
You are not so small as I,
And not half so spry.
I'll not deny you make
A very pretty squirrel track;
Talents differ; all is well and wisely put;
If I cannot carry forests on my back,
Neither can you crack a nut."

Animal House, or Life at Harvard College

What was life like at Harvard in the early 19th century? Here is a scene described by a student poet:

Nathan threw a piece of bread,
And hit Abijah on the head.
The wrathful Freshman, in a trice,
Sent back another bigger slice;
Which, being butter'd pretty well,
Made greasy work where'er it fell.
The fight was just beginning;
And thus arose a fearful battle;
The coffee cups and saucers rattle;
The bread bowls fly at woeful rate,
And break on many a learned pate.

Four students were suspended after the "Bread and Butter Riot." It was a foolish way to act, and hard on the dishes, but otherwise thought to be harmless. It didn't turn out that way for one Harvard student.

A Harvard freshman around 1850.

William Prescott was a junior when a classmate playfully threw a crust of bread at him. It cost him the sight of his left eye. Soon after his graduation, his right eye began to fail. But he didn't let that stop him. Prescott paid people to read to him. He believed history—real facts—could be as interesting as fiction (so do I). He wrote works of history that are accurate (he did careful research) and also exciting to read. His greatest success was *The Conquest of Mexico*, which used dramatic details to tell the story of the Spanish conquistador Cortés and the Aztec emperor Moctezuma. It is still good reading.

27 Mr. Thoreau— at Home With the World

I should not talk so much about myself if there were anybody else whom I knew as well.
—HENRY DAVID THOREAU, *WALDEN*

It is never too late to give up our prejudices.
—HENRY DAVID THOREAU, *WALDEN*

"My life has been the poem I would have writ," wrote Thoreau, "But I could not both live and utter it." What did he mean?

Some people say that of all Ticknor's students, Henry David Thoreau was the best thinker. But you certainly wouldn't have suspected that if you had seen him. He was small, thin, and muscular, with blue eyes, yellow hair, and a long nose like Ralph Waldo Emerson's. He wore old clothes. He liked people but he didn't do any extra socializing; he didn't have time to waste. He spent time on the things he thought important—like watching birds, or tracking a fox, or admiring a flower's blossom.

"Time is but the stream I go a-fishing in," wrote Thoreau. Now what did he mean by that?

Thoreau could do almost anything he wanted to do; he knew how to use his hands and his head. He was the son of a pencil maker, and sometimes he made pencils. He also milked cows and built his own house, a little cabin in the woods on the edge of Walden Pond. He borrowed some tools and chopped down some trees, and the house cost him $28.12. He lived in it for two years—from 1845 to 1847—and wrote a book about his stay there which he called *Walden*. That little book reads as if it was easy to write. It wasn't. Thoreau wrote and rewrote; he polished his words carefully. He was a craftsman with words.

Above man's aims his
nature rose.
The wisdom of a just
content
Made one small spot a
continent,
And tuned to poetry
Life's prose.
—LOUISA MAY ALCOTT ON HENRY DAVID THOREAU

WALDEN.

By HENRY D. THOREAU,

AUTHOR OF "A WEEK ON THE CONCORD AND MERRIMACK RIVERS."

You can visit Concord and see Walden Pond and Thoreau's cabin pretty much as they were when he lived there.

Rather than love, than money, than fame, give me truth, he wrote.

Most people spend their lives working so they may own things. Thoreau wished to own only himself.

> *Be a Columbus to whole new continents and worlds within you....explore the private sea, the Atlantic and Pacific Ocean of one's being.*

"No truer American existed than Thoreau," said Emerson.

The tax collector didn't think so. Thoreau thought the government was wrong for allowing slavery and for going to war in Mexico. He protested by not paying a government tax. Thoreau spent a night in jail and would have stayed longer, but a relative paid the tax. Thoreau thought each person can be important if he or she speaks out about injustice. He explained his thoughts in an essay called *Civil Disobedience.* That book has inspired people around the world to stand up for freedom and fairness in government. Thoreau believed in nonviolent action. (What does that mean?)

Thoreau loved nature. He looked at the world about him and learned from it. He wrote:

> *I tread in the steps of the fox that has gone before me....I know which way a mind wended, what horizon it faced, by the setting of these tracks, and whether it moved slowly or rapidly, by the greater or less intervals and distinctness; for the swiftest step leaves yet a lasting trace.*

"In Wildness is the preservation of the World," wrote Thoreau. Do you agree?

Two great 20th-century leaders, India's Mohandas (Mahatma) Gandhi and America's Martin Luther King, Jr., were inspired by Thoreau's ideas. Thoreau believed in nonviolent protest and in passive resistance. Passive resistance? That sounds like a contradiction (or an oxymoron), but it isn't. It means not fighting back when you are attacked, but not cooperating or running away or backing off either. That takes real courage.

This sketch isn't a typical one of Thoreau, who once said, "Beware of all enterprises that require new clothes."

28 Melville and Company

Oh, give me again the rover's life—the joy, the thrill, the whirl! Let me feel thee again, old sea! let me leap into thy saddle once more. I am sick of these terra-firma toils and cares; sick of the dust and reek of towns. Let me hear the clatter of hail-stones on icebergs, and not the dull tramp of these plodders, plodding their dull way from their cradles to their graves. Let me snuff thee up, sea breeze! and whinny in thy spray.

—HERMAN MELVILLE, *WHITE JACKET*, 1850

Reveries are daydreams. ***Remote*** means distant and ***barbarous*** means uncivilized.

A ***depression*** is a time of economic hardship, when businesses fail and jobs are scarce.

"We are the pioneers of the world," said Herman Melville of Americans.

Ralph Waldo Emerson and Henry David Thoreau went to Harvard. Herman Melville went to sea.

A whaleship was my Yale College and my Harvard, he wrote.

When he was a young boy Melville lived in New York. He often wandered onto the city's docks and daydreamed of voyages to faraway places:

> *I frequently fell into long reveries about distant voyages and travels, and thought how fine it would be, to be able to talk about remote and barbarous countries; with what reverence and wonder people would regard me, if I had just returned from the coast of Africa or New Zealand; how dark and romantic my sunburnt cheeks would look; how I would bring home with me foreign clothes of a rich fabric and princely make, and wear them up and down the streets, and how grocers' boys would turn back their heads to look at me, as I went by.*

The Melvilles were a wealthy family; then Herman's father died; and then, in the depression of 1837, the family lost its money. Melville needed a job, but there isn't much work to be found during a depression. Some people were seeking opportunity by heading west; Herman Melville signed aboard a sailing ship.

He went to the foreign places he had dreamed about. But sailing wasn't as glamorous a life as he had imagined.

> *Now I was a poor friendless boy, far away from my home, and voluntarily in the way of becoming a miserable sailor for life. And what made it more bitter to me, was to think of how well off were my*

cousins, who were happy and rich, and lived at home with my uncles and aunts, with no thought of going to sea for a living....I felt thrust out of the world.

In those days, gentlemen went to sea as officers; the life of a common sailor was hard.

When I go to sea, I go as a simple sailor, right before the mast, plumb down into the forecastle, aloft there to the royal mast-head. True, they rather order me about some, and make me jump from spar to spar, like a grasshopper in a May meadow. And at first, this sort of thing is unpleasant enough. It touches one's sense of honor, particularly if you come of an old established family in the land, the Van Rensselaers, or Randolphs, or Hardicanutes....But even this wears off in time.

Herman Melville's first voyage was on a merchant ship: to Liverpool, England. Then he went to sea on a whaling ship. As you know, whaling was important in those days before electric lights. People burned whale oil in the lamps that lit their homes. But whaling was dangerous; many whalers didn't return home to tell of it. Melville was one of the lucky ones.

Melville had seen all the excitement—and terror—of whaling.

Herman Melville sailed to England, to Cape Horn, to the Sandwich Islands (now called Hawaii), and to Tahiti, all before he was 25. At that age, in 1844, he was back on land—to stay. He began to write. *From my twenty-fifth year I date my life,* he said in a letter to Nathaniel Hawthorne. What do you think he meant by that?

The ***forecastle*** is the front section of a ship where the crew is housed. You can pronounce it FOR-kassel or FOKE-sul.

Melville wrote a book about a huge white whale named Moby-Dick, and about a ship's captain named Ahab who was obsessed with that whale—which means he couldn't get the sea creature out of his mind. That great white whale haunted Ahab; he was determined to capture it. Melville's *Moby-Dick* is the story of that obsession. It is a book of adventure, bravery, cruelty, and daring. It is a great American novel.

James Fenimore Cooper

But the most popular adventure stories of the 19th century were James Fenimore Cooper's *Leatherstocking Tales.* Their hero, Natty Bumppo, was a lot like Daniel Boone.

And there was Rip Van Winkle. Rip fell asleep—for 20 years. Washington Irving

Washington Irving

Rip Van Winkle comes home to a ruined house and tells his story to some skeptical-looking listeners. These pictures were drawn by one of the first artists to illustrate Washington Irving's story.

wrote about old Rip and about the Hudson River Valley in New York. Can you imagine what it would be like if you fell asleep and woke up 20 years from now? What would happen to your friends in that time? What about the place where you live? King George III was the ruler when Rip fell asleep. Do you know who was in charge when he woke up? (Read *The Legend of Sleepy Hollow* and you will find out.)

Herman Melville, Washington Irving, and James Fenimore Cooper had something in common: they were New Yorkers. Emerson, Thoreau, Alcott, and Hawthorne were New Englanders. Americans in every region were beginning to write.

Edgar Allan Poe was born in Boston but lived much of his life in the South. He wrote scary stories—very scary stories. Some people think he is the best horror writer of all time. He didn't have an easy life. His parents died, his wife died, and no one seemed to like his stories or his poems. It was only after he died—in 1849, when he was 40—that people began to appreciate his work. If you like good stories, and you don't mind being scared, read "The Tell-Tale Heart." While you're at it, try a poem that Poe wrote called "The Raven." It is especially good when read aloud.

Edgar Allan Poe

But if you want to read poems that sing with exuberance, try some of Walt Whitman's poems. Whitman was a big man, full of energy and good nature, who grew up in Brooklyn, New York, and became a teacher, a newspaper reporter, and an editor. He got fired from his newspaper job because of his antislavery ideas. That gave him time to do what he wanted, and what he wanted was to become a poet. Like Emerson and Thoreau, Whitman thought God could be found in people and in nature. "I was simmering, simmering, simmering, Emerson brought me to a boil," he said.

A long poem Walt Whitman wrote, called *Leaves of Grass*, was unlike any poem ever written before. At first no one much bothered to read it—and those who did didn't like it, because it doesn't rhyme, as much poetry does. Those who did read it found that it has rhythm and that it is a roaring, rollicking poem all about America, and about Whitman, and about ordinary things and extraordinary things. It is about a trapper, his Indian bride, a clipper ship, a runaway slave, a butcher-boy, a moose, a chickadee, and a prairie dog. It is about us.

No one knew what to make of Whitman's poem, except Ralph Waldo Emerson, who knew, right away, that Whitman was writing with a new kind of voice: an American voice. Emerson said *Leaves of Grass* was "the most extraordinary piece of wit and wisdom that America has yet contributed." Here at last was the American poet, the poet of democracy, that Professor Ticknor had hoped for.

In 1820, an Englishman, the Reverend Sydney Smith, was mocking America when he said, "Who reads an American book? or goes to an American play? or looks at an American picture or statue?"

By 1850, people all over the world were beginning to read American books—and to look at American paintings and statues, too.

Walt Whitman

Camerado, this is no book,
Who touches this touches a man.

—Walt Whitman, "So Long!", from *Leaves of Grass*, 1855

Tales That Are Whoppers Are Tall

Mike Fink

Some Americans were becoming great storytellers. They were developing an American specialty: the tall tale. European stories were often of privileged people: knights, princesses, kings, and queens. American tall tales were exaggerated stories about ordinary people who became heroes because of what they could do: like powerful Paul Bunyan and his blue ox, Babe; or Pecos Bill, a mythical cowboy; or Mike Fink, who was a real frontiersman and boatman, although most of the stories about him are tall tales. At least we think they are.

Paul Bunyan, who outdid everyone.

29 Painter of Birds and Painter of Indians

A Graphomaniac

In a lifetime that spanned the years from 1785 to 1851, John Audubon filled endless sheets of paper with his fine, flowing script. He seemed to write continuously, pouring out letters, journals, stories of the American frontier, notes on the scientific details of the birds and animals he studied. Sometimes he used a quill; later, he said, he "drove an iron pen." Occasionally he wrote so long and hard that his hand swelled, and for a time he would have to still the urge to spill his thoughts out onto paper. When he was away from home, he bombarded his family with letters five and six pages long, and kept up a voluminous correspondence with friends and colleagues. Such long, frequent letters probably weren't always welcome, since at the time the receiver had to pay for postage, by the page.

—FROM SHIRLEY STRESHINSKY, *AUDUBON*, VILLARD, 1993.

What happened in 1803 in the American West?

Audubon was so popular in England that his friend Sully wrote: "It is Mr. Audubon here and Mr. Audubon there until I am afraid poor Mr. Audubon is in danger of having his head turned."

His name was John James Audubon. He called himself an "American woodsman," which he was, although, like so many other Americans, he was also an immigrant who never lost his foreign accent.

Audubon was born in Santo Domingo (now Haiti) but grew up in France. His mother died when he was a baby, but his father and stepmother loved him dearly. His stepmother told him he was the handsomest boy in all of France. His parents tried to give him everything he wanted. Later, he said that perhaps they shouldn't have spoiled him so, but his childhood was a happy one.

He loved to roam the out-of-doors, and he filled his bedroom with birds' eggs and nests and snakeskins; then he drew pictures of all the things he brought in from the woods. He drew with talent, but no one believed he could earn a living with his drawings. He was expected to go to sea, as his father had. But when he did he got seasick.

Then Napoleon came to power and needed soldiers. The young artist knew if he stayed in France he would be drafted into the army. So in 1803, at age 18, Audubon came to America to escape Napoleon's war. His father, who had fought with Lafayette in the American Revolutionary War, owned a farm near Philadelphia. That farm, called Mill Grove, was managed by a sensible Quaker couple. Audubon's parents thought they would be a good influence on their handsome son. Young Audubon was charming and high-spirited; in France he had spent most of his time dancing, playing the violin, fencing, drawing, and wandering

John James Audubon: The Nature of America

Audubon's birds and animals, clockwise from top: the great white heron; the ferruginous mocking bird (fending off a snake); and the nine-banded armadillo.

in the woods. It was time for him to get serious, his parents said.

Guess what Audubon did at the farm in America? He danced, drew pictures, made music, wore fancy clothes, and went wandering in the woods. Whatever he did, people couldn't help but like him: he was always good-natured and optimistic.

Then two things happened that changed John James Audubon's life: he became engaged to Lucy Bakewell, and his parents were no longer able to support him. How would he earn a living? He decided to head out to frontier Kentucky and open a store for settlers who would surely need supplies. He and his business partner climbed aboard a big flatboat and went west on the Ohio River, sharing the deck with horses, cattle, pigs, trappers, hunters, and families with children. Because they feared pirates they traveled only at night—without lights. During the day they tied the boat to the shore, built campfires, told stories, and danced and sang on the deck. Audubon played his flute.

Audubon opened his store in Louisville. A year later he returned east, married Lucy, and brought her to Kentucky. Soon they had two sons: Victor and John. But Audubon wasn't very good at shopkeeping.

From New Orleans, Audubon wrote: "I took a walk with my Gun this afternoon to see...Millions of Golden Plovers Coming from the North East and going Nearly South—the destruction...was really astonishing—the Sportsmen are here more numerous and at the same time more expert at shooting on the Wing than anywhere in the U. States." Audubon said there were about 400 hunters that day and each shot about 30 dozen birds. You can figure out the total number killed in one day.

Picture Plates

An engraving is a picture printed from a series of metal plates. The original drawing must first be copied onto the plates. Acid is used to cut the lines of the drawing into the metal. Then colored ink is rubbed on the plate. The ink runs into the cut lines. The rest is wiped away. Paper is pressed against the plate and the ink makes a print. A separate plate is needed for each color ink. The paper is pressed onto the plates, one after the other. Colored drawings, like Audubon's, need many inkings and many plates. A black-and-white engraving needs only one plate.

Audubon drew the natural world around his animals as carefully as he drew the animals themselves. Here the chestnut-collared lark bunting collars a spider.

He kept drawing, or going off into the woods to hunt. He wore buckskin clothes and moccasins, carried a tomahawk, and, with his flowing long hair and deep-set eyes, looked like an Indian. His biographer said that "Audubon often stayed in Indian camps or went hunting with braves; he believed Indians to be a heroic people, and he admired their simplicity and modesty."

One day a famous bird expert—an ornithologist—came into his store. He was Alexander Wilson and he showed Audubon his drawings of birds. He said they were the best bird drawings ever done in America, and they were. But as soon as Audubon saw them he knew he could do better. Wilson's pictures were of stuffed dead birds. Audubon was drawing birds that were doing things—feeding their young, or building nests. Audubon drew his birds in a real, natural world.

After Audubon saw Wilson's drawings he got serious about his own. Soon it was no longer a hobby; he made it his life's work. He decided he would try to draw all the birds of North America. He was in a good place to do it: vast flocks of birds migrate through the Mississippi Valley. He knew he had to hurry. Hunters were killing many birds; settlers were cutting down trees and destroying nests; some of the birds would soon be extinct.

Audubon went into the woods and watched the way birds lived and what they ate. Sometimes he caught a bird, put it in a cage, drew it, and then set it free. Sometimes he shot birds and used wires to make them look as if they were alive. He would lay a bird on a paper marked with big squares, and then take another piece of paper with the same size squares and copy the bird. He drew birds life-size and he made the size exact.

But people on the frontier didn't need bird drawings. Audubon had to do other things to earn a living. He taught dancing; he taught drawing; he hunted. Lucy went to work as a teacher. It was a discouraging time. "My best friends solemnly regard me as a madman," he wrote in his journal. But he was perfecting his talent as an artist. He was also becoming a scientist. As far as we know, he was the first person to put a band on a bird's leg and then wait to see if it would return in the spring after flying away for winter. (It did.)

Audubon went to Philadelphia to look for a publisher to print his drawings. But no one in America seemed to have the skills to do it as carefully as Audubon wished. Lucy had saved some money; she sent it to her husband so he could take his drawings to England. In Liverpool and London he was a great sensation.

Everyone was enchanted with the American woodsman. He wore a fur cap and a wolfskin coat and he shined his shoulder-length hair with bear grease (as Daniel Boone did). In England people paid to see his drawings. An English engraver was amazed. "I never saw anything like this before," he said. "Who would have expected such things from the wilds of America?"

Still, it wasn't easy getting the drawings printed. Audubon was a perfectionist who insisted on the best-quality work. He was lucky: in London he found a brilliant engraver. It took 11 years to engrave and print *Birds of America*. While that was happening Audubon kept on drawing. He knew that some of the birds and animals he drew might soon disappear from the planet.

George Catlin felt the same way about the people whom he painted. When Catlin's mother was eight she was captured by Iroquois. The little girl, who had been living on a frontier farm, was taken to a longhouse, treated well, and later released. She told her son stories of that adventure, and he never forgot them. When he grew up he became a member of the

George Catlin: Student of the American Indian

Long before almost anyone else, George Catlin understood that both Indian and buffalo were in danger of extinction. In 1832, when he painted this Sioux Buffalo Chase, *he called on his countrymen to save the Great Plains as a national park and Indian reserve.*

The bear-claw necklace of White Cloud, Head Chief of the Iowas *(left), was a sign of bravery and great importance. Right, Catlin sketched himself at work, painting a Mandan chief.*

The United States of Artists

Gilbert Stuart

The Revolutionary War period produced America's first group of fine artists. Most people agree that the greatest painter of that time was John Singleton Copley of Boston. Copley taught himself to paint and then did portraits of people who seem so real you can almost tell what they're thinking.

Gilbert Stuart, from Rhode Island, was another outstanding portrait painter. You've seen one of his paintings. You've seen it many times. It is the portrait of George Washington on every dollar bill.

And then there was that extraordinary museum keeper, Charles Willson Peale, who painted 60 portraits of George Washington. He did more than 1,000 other portraits of people who included Martha Washington, Benjamin Franklin, Thomas Jefferson, and John Adams. (Remember, there were no cameras then. If you wanted to have your picture made, you hired a portrait painter.)

In 1822 Charles Willson Peale painted himself and his museum, which had stuffed animals and fossils as well as art.

Peale was versatile. He was also a saddlemaker, a watchmaker, a silversmith, and an inventor. He took a set of elk's teeth and made them into dentures for George Washington. He invented a velocipede—which was a kind of bicycle. He worked with Jefferson on a polygraph (not a lie detector—a ma-

All of Charles Willson Peale's family were named for famous artists. Here, Rembrandt Peale painted his brother Rubens.

Pennsylvania Academy of Fine Arts, and, before long, was painting portraits of wealthy folk. But he wasn't happy. He had a yearning to do more with his talent than paint pretty portraits. Maybe it was his mother's stories that made him seek out Indians and sketch and paint them. They were hard to find. Most of the Eastern Indians were dead or had been moved west.

In 1830, George Catlin decided to find and paint Indians living in their traditional ways. So he went to St. Louis, which had become the gateway to the West.

Then he had some colossal luck. Or maybe it was just good sense. He chine that copied what you wrote down as you wrote it). And he encouraged young artists, like John James Audubon.

When he was in Philadelphia, Audubon visited Charles Willson Peale's museum on the second floor of Independence Hall. Peale's museum was said to be the first in the United States. Inside were Indian relics, the skeleton of a mastodon, wax dummies, paintings—anything Peale found interesting. Peale invented his own way of stuffing animals. Then he displayed the animals in a natural environment—it was 100 years before most other museums used that idea. Peale fathered 17 children and named most of them for famous artists. Raphaelle, Rembrandt, Titian (who went as an official artist on Stephen Long's expedition), and Rubens Peale all became artists themselves. Joshua Johnston, a free black man who lived in the Peale household, was also a skilled portrait artist. Europeans were astounded when they met John Singleton Copley, Charles Willson Peale, and Gilbert Stuart. They seemed marvelous examples of the talent democracy could produce.

Joshua Johnston's Young Lady on a Red Sofa. *Few black people, of course, could pay to be painted.*

John Singleton Copley's famous painting of Watson and the Shark *in Havana harbor. Watson, an English midshipman, was rescued—minus a leg.*

looked up William Clark—of the Lewis and Clark expedition—and told him what he wanted to do. Clark was 60, and Superintendent of Indian Affairs. Catlin was 34 and eager. They got along splendidly, and, because the young man was sensitive and genuine, Clark taught him what he knew.

In 1832 Catlin took off, on the first voyage of the steamboat *Yellowstone* up the Missouri River. When he returned, a year later, he had sketchbooks filled with paintings of the Mandan, Sioux, Blackfoot, and Crow Indians. He had painted the dance that celebrates the buffalo hunt, and sketched a torture ceremony intended to select the bravest of a

I love a people who have always made me welcome to the best they had...who are honest without laws...who have no poor house....who never raised a hand against me or stole my property... and oh! how I love a people who don't live for the love of money. —GEORGE CATLIN ON INDIANS

Samuel Morse was born into a well-known family (his father, Jedidiah Morse, wrote the first American geography book). He was an artist as well as an inventor, and he painted this self-portrait in 1811, when he was 20 years old.

tribe's young men. He had drawings of clothing, tattoos, bear-claw necklaces, ball games, tents, and people. At a time when many Americans were ignoring—or attacking—Native Americans, George Catlin was painting them with honesty and affection.

Audubon and Catlin weren't America's only fine artists. In those days, artists were record-keepers. There was no other way to have a picture of yourself, or of your home, or of scientific data. So artists, especially portrait artists, were in demand. Philadelphia artist (and friend of Audubon) Thomas Sully was known for his fine manners and his graceful portraits; Samuel F. B. Morse, who invented the telegraph, was a portrait artist, too; William Sidney Mount painted Americans in action—farming, dancing, and listening to politicians; and George Caleb Bingham, from Missouri, painted boatmen and trappers and frontier Americans. Hiram Powers, Thomas Cole, Edward Hicks, and Asher B. Durand were other 19th-century American artists.

Was American art worth considering? Or do you think the Reverend Sydney Smith was right?

Bingham and Mount: America at Work

In 1845 George Caleb Bingham (above, in a self-portrait) painted the solitude of these Fur Traders Descending the Mississippi.

The excitement of Ringing the Pig *(1842), by William Sidney Mount (left, in an 1832 self-portrait).*

30 *Amistad* means Friendship

In Havana, Cuba, where Sengbe Pieh was sold after being captured illegally, the Spanish gave him a certificate calling him a Ladino: a Spanish-speaking slave not born in Africa. Such slaves could still be sold.

In 1839, Sengbe Pieh was working on a road connecting his small mountain village to the next tribal town. He was 25 years old, with shining brown skin, close-cut black hair, clear wide eyes, and a face that reflected a nature both innocent and strong as a nut tree. Pieh was not a prince—as some would later call him—but he was a born leader. He was the father of three children.

Pieh had never been far from his village in Sierra Leone, but when four strange men stepped out of the bush and surrounded him, he knew at once who they were. Sierra Leone, in Africa, was a British colony. It had been a center of the slave trade for almost 300 years; since 1562 to be exact, when England's Sir John Hawkins set out on a slave-gathering expedition.

Slave trading had made people rich: in England, the United States, Spain, the Arab nations, and Africa. But now there were abolitionists, who disapproved of slavery. They were horrified by the idea of selling people, and they were doing something about it. In 1787, some British abolitionists founded Freedom Province, on the Sierra Leone coast. In 1792 they built the city of Freetown and began returning blacks to Africa.

In 1808 the United States outlawed the foreign slave trade. That means it became a crime to bring new slaves into the country. Those who were already slaves could still be bought and sold inside the United States. Be sure that you understand that it was the slave trade—not slavery itself—that was outlawed. No more Africans were to be brought to the United States. (Britain and Spain also made foreign slave trading illegal.) The

I teach the kings about their ancestors so that the lives of the ancients might serve them as an example, for the world is old but the future springs from the past.

—Mamadou Kouyaté, a Mali griot, from *Sundista: An Epic of Old Mali* (1217–1237)

slavers—those who ran slave ships—were now criminals. The penalty, if a slaver was caught, was death.

So in 1839 Pieh should have been safe. But he wasn't. It may have been illegal, but there was still big money to be made by selling slaves. (Just as, a century or so later, there would be big money made selling illegal enslaving drugs.)

Yoked in a *coffle* or neck fetter (coffle comes from the Arabic word for caravan, *cafila*), these West Africans are marched to the Bight of Benin on the Atlantic. There they are kept tied up in the *barracoon* or slave barrack until the ship has a full load. Below is a painting of a slave ship's hold (made after the ship was captured by an antislavery ship).

The four men would not let Pieh say good-bye to his wife and children. Pieh wondered if they would know what happened to him. Would they think he was eaten by a lion, or would they guess the truth?

The rough men with guns had captured other Sierra Leoneans. For three and a half days they kept them marching. Finally they came to the coast. Most had not seen the ocean before; the roaring water terrified some of them.

There were worse terrors to come. They were soon to be chained—neck to neck and ankle to ankle—and thrown inside a ship into a hold so low they could only sit, not stand. For two months they would be held like that—with rice to eat but little water. One in every three of them died. Finally they were brought on deck, allowed to wash, and given some extra food. Pieh knew the journey must be coming to an end. But where were they? What would happen next? He heard the word *Cuba*, but he didn't understand Portuguese—the language of the ship's crew—and he knew no one would answer his questions anyway.

And then the ship stopped. The crew put spyglasses to their eyes and searched the water nervously. The slavers were watching for British cruisers. They were doing something illegal and they knew it. If they were captured no one would come to their rescue. Nightfall came and quietly the ship proceed-

A Natural Variety of Classes

I repudiate, as ridiculously absurd, that much lauded but nowhere accredited dogma of Mr. Jefferson, that "all men are born equal." No society has ever yet existed...without a natural variety of classes....Slavery is truly the "corner-stone" and foundation of every well-designed and durable "republican edifice," ***wrote slave owner James Henry Hammond from his home at Silver Bluff, South Carolina, in 1845. Hammond's letter was addressed to an abolitionist. He continued:***

You will probably say, emancipate your slaves, and then you will have free labor on suitable terms. That might be if there were five hundred where there now is one, and the continent, from the Atlantic to the Pacific, was as densely populated as your Island. But until that comes to pass, no labor can be procured in America on terms you have it....I have no hesitation in saying that our slaveholders are kind masters, as men usually are kind husbands...and friends—as a general rule, kinder.

What is wrong with Hammond's argument? Whose opinion is he ignoring?

ed. The Africans were loaded into small boats, rowed ashore, marched into the Cuban jungle, and put in crude shacks. The slavers had made it.

Now they needed to show that these Africans were actually slaves (or children of slaves) brought to the island before the slave trade was made illegal. So each African was given a European name and false identity papers. Sengbe Pieh became Joseph Cinque (SIN-kway). Once that was done the Africans were ready to be sold on the legal market.

On board ship, many slaves got sick: cleanliness was almost impossible, the food was scanty and bad, and they were often beaten. Many sickened from depression and terror, or from being kept locked up below decks. Slavers forced them to dance to fife or drum, to keep their muscles in shape. It was another horrible humiliation.

And that is exactly what happened. They were marched to Havana, the Cuban capital, and put in a big, open-air stockade. Buyers came and looked them over. There was one problem that everyone ignored. None of these Africans could speak Spanish. They could speak only African languages. Some were children. Anyone with sense could guess they had not been born in Cuba. But, as I said, there was big money in this business. Cuba was a Spanish colony. Both the Spanish and the Cuban authorities looked the other way.

José Ruíz went into the stockade looking for slaves to buy. He saw the newly arrived men. He noticed Cinque. He examined his teeth. He was pleased. Joseph Cinque was lean, muscular, and healthy. Ruíz paid the captain $450 for Cinque and the same amount for each of 49 other blacks. His partner, Pedro Montes, bought four children: three girls and a boy. The oldest was nine.

If the Africans thought they were through with ships they were wrong. They were now dragged on board a small Baltimore-built schooner, the

This is an artist's version of the *Amistad* story. But it doesn't make sense that the ship's crew would stand around watching while the slaves are roused to mutiny. Perhaps Cinque is telling his companions what has happened to them since they reached America.

A ***mutiny*** is a rebellion against authority, especially when sailors or soldiers rebel against their officers.

Amistad. They were headed for an island port. The captain, four crew members, and Ruíz and Montes were breaking the law—they knew these people had not been born in Cuba—but they weren't worried. Their papers seemed legal. They had done this before.

Each black prisoner was given a banana, two potatoes, and a small cup of water—as a day's rations. It was hot and sultry. When one man took extra water he was beaten, and gunpowder was rubbed in his wounds.

Cinque was terrified. Using sign language, he asked the brown-skinned cook, Celestino, what was going to happen to them. The cook—in a cruel joke—ran his finger across his throat. Then Celestino pointed to barrels of beef—they were part of the ship's cargo. He laughed. Cinque thought he was to be turned into meat. Were Ruíz and Montes cannibals? They were cruel and crude enough. Cinque shuddered at the thought.

Then he found a big iron nail. Quietly Cinque used the nail to pick the lock on the chain that circled his neck. That night he freed the others. Frantically they looked for weapons and found boxes of sugar-cane knives—knives with fierce blades two feet long.

It was stormy that night, and dark, without moonbeams. The crew had lowered the sails. The ship was quiet. The mutiny began at 4 A.M. Fifteen minutes later it was over. The captain was dead, and so was Celestino. Two sailors had jumped overboard and were not heard of again. Ruíz and Montes were prisoners. Antonio, the captain's slave, was unhurt.

Now the problem Cinque faced was how to get back to Africa. He didn't know how to sail; he'd hardly been on deck before. He didn't know where he was. He did know that Africa lay in the direction of the rising sun. So he forced the two Spaniards to sail east—toward the sun—and prayed they would make it across the ocean.

What he didn't know was that every night Ruíz and Montes changed course and sailed back west. So the *Amistad* zigzagged along the American coast heading for New York, not Sierra Leone.

At last they spotted land, but it was to the west. Cinque knew he had been tricked. He also knew he needed food and water. He took a rowboat to land and was soon captured. The *Amistad*, the Africans, the two Spaniards, and the four children were in Connecticut waters. Now what was to be done with them?

In 1839, President Martin Van Buren was planning to run for re-election. The last thing he wanted was an incident that would cause controversy. Anything that had to do with slavery was guaranteed to do that. The *Amistad* couldn't have landed at a worse time—for Van Buren.

This was the situation:

Blacks had mutinied and killed whites. That was just what every Southern plantation owner feared.

A slave ship had been captured with blacks aboard who were courageous enough to fight for their freedom. That was just what the Northern abolitionists had been waiting for.

A Spanish ship had been captured. The Spanish ambassador called it an "outrage."

A slave ship had broken the terms of the Spanish-English antislavery treaty. Would the British use that as an excuse to invade Cuba? Clearly, this was no simple affair.

"Send the blacks back to Cuba," said those who approved of slavery.

"Set them free," said those who didn't.

This matter would have to be decided in the courts. In Connecticut a judge said the Africans had been captured illegally. The laws prohibiting the slave trade had been broken, he said, and the captives should be freed.

That wasn't the end of it. Everyone knew that the Connecticut judge hated slavery. The case was appealed. The next judge was a New

Martin Van Buren lost the race for president in 1840 when he ran against "Tippecanoe and Tyler Too." That was the campaign slogan of Whig William Henry Harrison (who fought Native Americans at Tippecanoe) and his vice president, John Tyler.

There's no mistaking this for anything but the mutiny: the blades of those knives are three inches wide at the ends.

Old Man Eloquent

When John Quincy Adams entered the House of Representatives, in 1831, he was 64, short, stout, bald, with a sense of duty that often made him seem quarrelsome. He was an ex-president; he was brilliant; but he saw himself as a failure—until he became a congressman. He loved that job, was reelected eight times, and fought—with all his might—against slavery. He knew that was important. The supporters of slavery in Congress had insisted on a "gag rule." That meant that any antislavery petitions were laid aside without discussion. The Southerners didn't even want the subject brought up. Adams kept introducing antislavery petitions anyway. He fought the gag rule—doggedly, stubbornly, ceaselessly. Finally, in 1844, the rule was defeated. In 1848, Adams had a stroke and died right in the House of Representatives. Even his enemies knew that an uncommon man was gone.

Making Up History

You will read in some books that Cinque returned to Africa and became a slave trader himself. That is not true. And yet that story has been written many times. Why? Because an author who learned the story of the *Amistad* and Joseph Cinque decided to write a novel about it. A novelist can write anything that makes a good story. He decided it would give the story an ironic twist to have Cinque become a slaver himself. A historian read the novel, thought it was true, and retold the story in a history book. (History books, of course, should always be true.) Then another historian quoted the first historian, and then another, and another. And that is how madeup stories sometimes come to be history.

Englander with a different record. He seemed to approve of slavery. He had made a teacher, Prudence Crandall, close the school where she taught black students. The abolitionists were discouraged.

President Van Buren had a ship readied in the New Haven, Connecticut, harbor. He thought the *Amistad* affair would soon be over. The ship was to take the blacks back to Cuba as soon as the decision was announced. Van Buren wanted them out of the United States—quickly—before another appeal could be made.

The judge studied the law. He may have approved of slavery, but he knew the law was more important than his personal feelings. There was no question about it, he said. These were free people—captured illegally. They couldn't be taken to Cuba against their will.

Those who believed in slavery were furious. They had only one place left to go—the U.S. Supreme Court, where five of the nine Supreme Court justices were Southerners. Now the abolitionists were desperate. So were Cinque and the men and the children. They had learned some English. They understood what was going on. Besides, they were weary and wanted to go home. The abolitionists asked an old man to help them. He agreed and said he would take no money for doing it. His name was John Quincy Adams. Some people called him "Old Man Eloquent."

In 1841, Adams stood before the Supreme Court and talked for three hours. One of the justices said it was an "extraordinary" argument. It all came down to one thing, said the former president, and that could be found in the Declaration of Independence.

> *I know of no other law that reaches the case of my clients, but the law of Nature and of Nature's God on which our fathers placed our own national existence.*

The Supreme Court agreed.

Cinque and his companions were free.

31 Webster Defends the Union

"I owe a paramount allegiance to the whole Union," says Clay to the Senate, "a subordinate one to my own State."

Secession seen from the North.

The year is 1850. The century's three most eminent senators—Henry Clay, John C. Calhoun, and Daniel Webster—are old men. All will soon be dead. But they will not go quietly. They are about to play their last scene in the national drama, and it will be a knockout performance. The country is being pulled apart—everyone can see that. Each time a new state enters the Union, the balance in Congress between North and South is threatened. Now California wants to become a state. California's constitution prohibits slavery. If California enters the Union, free states will outnumber slave states. Suppose the free states pass a law outlawing slavery?

John Calhoun, from South Carolina, cannot imagine the South without slavery. Calhoun calls slavery a "positive good." He says slavery is good for slaves and masters. He really seems to believe that. He is trying to unite the South. Calhoun says that if a state thinks a law is unconstitutional, it has the right to "nullify," or not obey, that law. He even says each state has the right to leave the Union if it wishes. That act of leaving the Union is called "secession." Some Southerners are beginning to talk of secession.

But other Southerners don't agree with Calhoun's states' rights argument. They don't like the idea of secession, even though they like slavery. They are Unionists. They believe in the United States.

Some white Southerners hate slavery. Virginia's Robert E. Lee has made that clear in letters to his family—but he has not spoken out in public. Kentucky's Cassius Marcellus Clay has. He is Henry Clay's cousin, and he has freed his slaves and taken abuse and even gunfire for his beliefs.

"I never use the word 'Nation' in speaking of the United States," says John Calhoun. "We are not a Nation, but a Union, a confederacy of equal and sovereign states."

Cassius Marcellus Clay

Slavery is ruining the South, says Cassius Clay. He says the slaves should be freed gradually and the slave owners paid for their financial loss.

It is not a bad idea, but most abolitionists won't even consider it.

Calhoun won't consider it either; he will have nothing but a slave-based society. He is a powerful thinker who has convinced most white Southerners with his arguments.

But he hasn't convinced Henry Clay. Clay, another Kentuckian, has been working on a compromise. As soon as people hear he is to speak, they pack the Senate chamber. Clay is 73 years old and white-haired, but still a commanding figure. He is wearing a black suit and a stiff white collar that touches his ears. Clay has a bad cough, but that does not keep him from speaking clearly and eloquently. He pleads for tolerance and understanding. Then he introduces his compromise. (Today it is known as the Compromise of 1850.) This is what it says:

1. *California is to be admitted to the Union as a free state.*
2. *New Mexico and Utah will become territories. (No mention is made of slavery; it is assumed the territories will decide that for themselves.)*
3. *A fugitive slave law will be enforced. (That means that runaway slaves who make it to free states must be returned to their owners.)*
4. *Slaves may no longer be bought and sold in the nation's capital. Slavery, however, will still be legal in the District of Columbia.*

John Calhoun is too sick to attend the session, but he learns all the details of Clay's speech. A few weeks later he comes to the Senate to answer Clay. He is wrapped in a great cloak and needs help to reach his seat. His face is so pale that he looks like a ghost. His dark eyes are feverish. Calhoun is too ill to speak; his speech must be read for him.

Clay thought he had given Calhoun what he wanted: a fugitive slave law. But Calhoun is still not satisfied. The North must stop talking about slavery, it must "cease the agitation on the slave question." If the abolitionists are not silenced,

> *let the States...agree to separate....If you are unwilling we should part in peace, tell us so, and we shall know what to do.*

A month later, John C. Calhoun is dead.

The debate over the compromise is not over. It is now Daniel Webster's turn. He will give one of the most famous speeches in Senate history. It is Webster's last Senate oration. Perhaps he knows that. Certainly everyone else does. There isn't an empty seat in the Senate chamber. "I wish to speak today, not as a Massachusetts man, nor as a Northern man, but as an American," says the mighty Daniel Webster. "I speak today for the preservation of the Union. Hear me for my cause." Webster has always

Party Time

It was the middle of the century, and the two big parties, the Whigs and the Democrats, were split inside themselves over slavery and the growing conflict between North and South. The Free Soil Party—which stood against the idea of bringing slavery to the western territories—just couldn't seem to find a strong presidential candidate. And the men who got elected—Millard Fillmore (Whig) and then Franklin Pierce (Democrat)—were, at best, mediocre. A new party was certainly needed. So, in 1854, some political leaders met in Ripon, Wisconsin. The Republican party developed out of that meeting. That same year, the Republicans won 100 seats in Congress. The party stood against slavery in the American West—but not for the moral reasons that led the abolitionists. At first the party's main appeal was to free white workingmen who didn't want to compete with slave labor. Six years after its founding, the Republicans put a president in the White House. His name was Abraham Lincoln.

been Clay's opponent—now he is agreeing with Henry Clay! People gasp. But Webster will do almost anything to save the Union.

> *I hear with distress, and anguish the word "secession," especially when it falls from the lips of those who are eminently patriotic, and known to the country, and known all over the world, for their political services. Secession! Peaceable secession! Sir, your eyes and mine are never destined to see that miracle.... There can be no such thing as a peaceable secession. Peaceable secession is an utter impossibility. Is the great Constitution under which we live, covering this whole country; is it to be thawed and melted away by secession, as the snows on the mountain melt under the influence of a vernal sun—disappear almost unobserved and run off? No, sir! No, sir! I will not state what might produce the disruption of the Union; but, sir, I see it as plainly as I see the sun in heaven. What that disruption must produce...such a war as I will not describe.*

To avoid that war, Webster will agree to a compromise. He believes in the Union and the American system:

> *We have a great, popular, constitutional government, guarded by law and by judicature, and defended by the affections of the whole people.*

When Webster finishes his oration some people weep. Is it because they know the Union is falling apart? Or is it because their hero, the "Godlike Daniel," has agreed to a fugitive slave law?

But Webster knows that if the South secedes now, the North will not be able to stop it. There will be two nations—one slave, one free—and perhaps wars between them. His speech does what it was meant to do. It holds the Union together. Congress votes to accept Henry Clay's compromise. The real problem is that no one knows how to end slavery, and still hold North and South together.

"I shall know but one country. The ends I aim at shall be my country's, my God's, and Truth's," says Daniel Webster. "I was born an American; I will live an American; I shall die an American."

Both Henry Clay and Daniel Webster die in 1852.

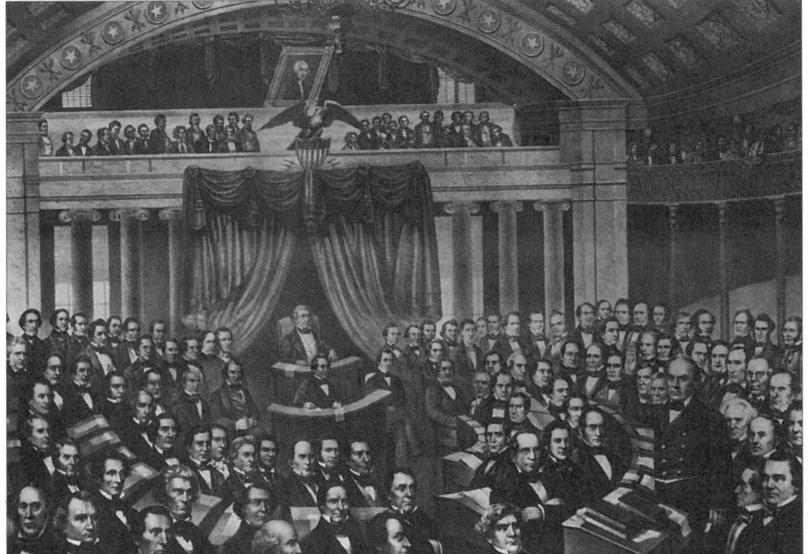

Daniel Webster gives his last great oration in Congress to a packed Senate chamber.

32 Big Problems and a Little Giant

When Franklin Pierce was inaugurated as president in 1853, he spoke of slavery, saying, "I fervently hope that the question is at rest." He must have thought that the Clay–Webster–Calhoun compromise of 1850 had solved the problem of slavery; but it certainly hadn't, and the president of the United States didn't even know it!

"I have taken a through ticket and checked all my baggage," said Stephen Douglas. He was talking about his commitment to "popular sovereignty" as a way of deciding the slavery issue in the territories.

Senator Stephen A. Douglas was known as the Little Giant. He was just over five feet tall, but so full of energy he was called a "steam engine in breeches." He'd been born in Vermont, but he moved to Illinois, where he became rich and famous as a political leader, a lawyer, and a businessman. He made his fortune in land speculation and in that new enterprise: railroads.

Senator Douglas could see that railroads were the future and that they would someday stretch from coast to coast. But what route would the transcontinental railroad follow? If it went from Chicago to San Francisco, Douglas's Chicago property would become even more valuable than it was.

There was something that stood in the way of that plan. It was the Indian territory west of Missouri, Iowa, and Minnesota. It was land left from the Louisiana Purchase, and most of it had been guaranteed to the Indians "as long as grass shall grow and water run."

The South was convinced that the North wanted to squash it—economically and politically. The answer to their problem was more land and power. Where could they find it? By controlling the western territories and making them slave territories.

Well, Indian treaties had been broken before—especially when there was money to be made—so that didn't seem a major obstacle, although a few people, like Sam Houston of Texas, would fuss about it. No, the problem was that Southern congressmen, such as Jefferson Davis, wanted the transcontinental railroad to take a Southern route, through New Mexico and Arizona.

Senator Douglas introduced a bill in the Senate to organize that

Louisiana Purchase land as the Nebraska Territory. Settlers (and land speculators) would be allowed to buy land, and, as soon as there were enough people living there, the territory would turn into states. According to the terms of the Missouri Compromise, they would be free states. Once there were settlers, the railroad trains would follow.

But that bill wasn't going to pass Congress—not as long as the South had votes. If Stephen Douglas really wanted his railroad there was a way he could have it. He could write a new bill that did away with the Missouri Compromise and opened the territory to slavery. Then he'd have his votes, said David Atchison, who was a senator from the slave state of Missouri. Otherwise, said Atchison, he would see Nebraska "sink in hell" rather than organize it as free soil.

Stephen Douglas introduced a new bill. This one divided the territory into two regions. Both had Indian names: *Kansas* and *Nebraska* (the regions were much bigger than today's two states with those names). The Missouri Compromise was repealed: the ban on slavery was ended. Instead, the residents of the territories were to decide for themselves about slavery (like the residents of the former Mexican territories). That idea was called *popular sovereignty*, and it seemed a democratic solution to Douglas.

Douglas wasn't a supporter of slavery. He didn't even like slavery. But he wasn't a slave, and he didn't worry about those who were. And he completely misjudged the feelings of people in the North. The Kansas–Nebraska Act "may have been the most important single event pushing the nation

Haste Makes Waste

Stephen Douglas didn't think carefully about what he was doing when he repealed the Missouri Compromise. Three years later he did realize the consequences when he saw some of the Southern extremists in action. This is what he said in 1857:

I was ashamed I had ever been caught in such a company. They are a set of unprincipled demagogues, bent on perpetuating slavery, and by the exercise of that unequal and fair power, to control the government or break up the Union.

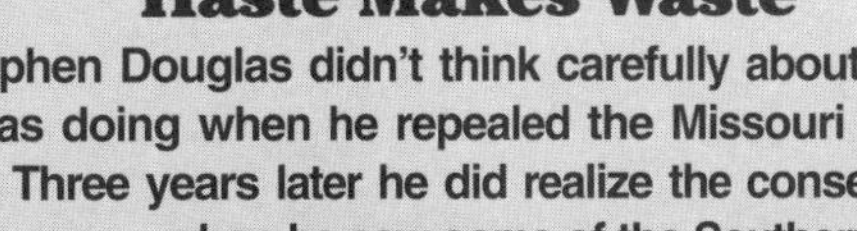

This cartoon shows Douglas hatching out a nest of problems: he holds an egg labeled "Missouri Compromise."

The proslavery and antislavery factions in Kansas Territory quickly moved from mass meetings to open warfare.

The Missouri Compromise, passed back in 1820, said that (1) Missouri was to be a slave state; (2) Maine (until then part of Massachusetts) was to be a free state; and (3) there was to be no slavery in the Louisiana Territory north of latitude 36°30', except in Missouri.

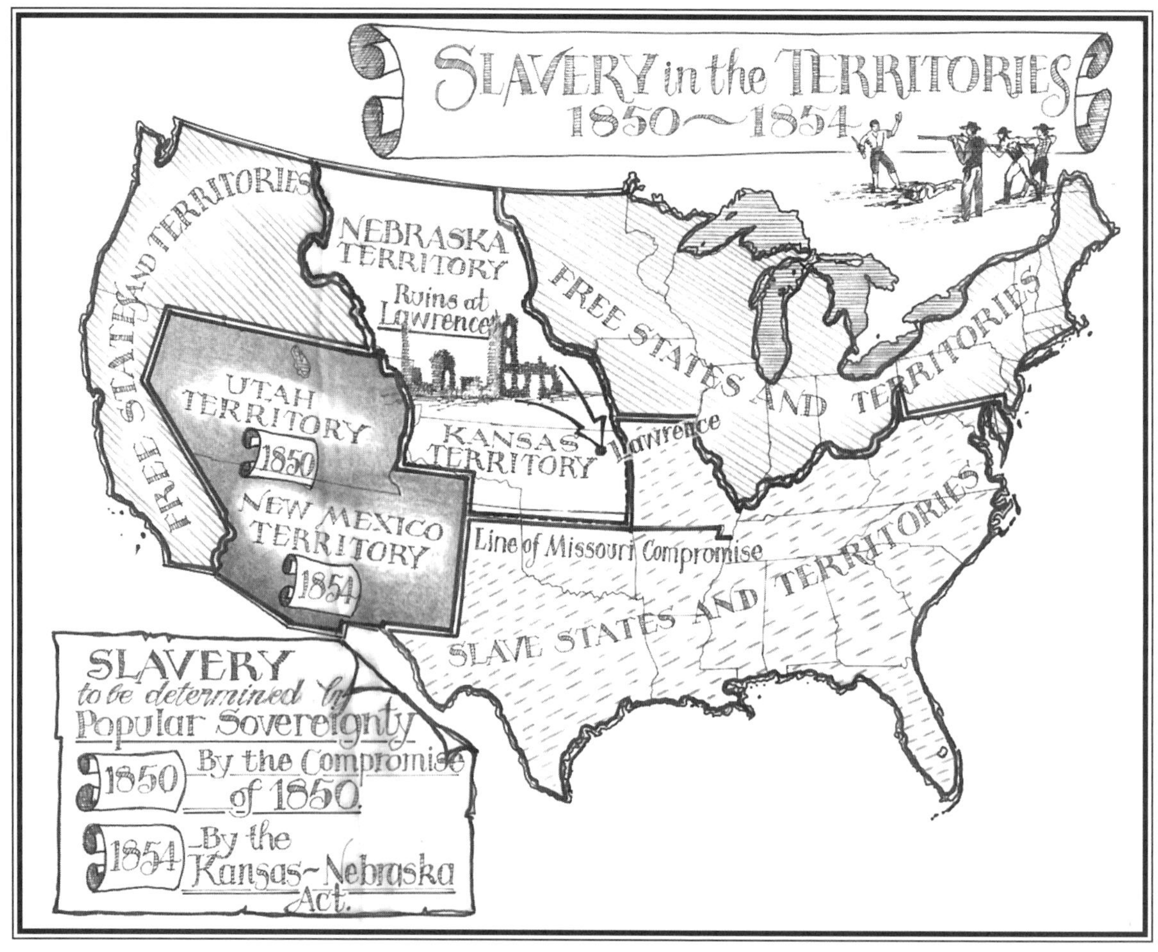

toward civil war," says historian James McPherson.

It made most Northerners furious. Henry Clay's Missouri Compromise had kept the peace between North and South for 34 years—since 1820. It was a link to the founding generation. It seemed almost sacred to many Americans. The Founders hadn't approved of slavery—they'd made that clear in a Declaration that said *All men are created equal.* They'd taken first steps to end slavery: they had abolished the slave trade. They had prohibited slavery in the Northwest Territory. They had talked about gradual emancipation.

The Kansas–Nebraska Act was a break with that tradition. If the Missouri Compromise was abandoned, what would be next? It was clear that Southern slave owners wanted to make the whole nation accept slavery. All at once there was a torrent of fiery speeches, editorials, and sermons in the North—all were against this Kansas–Nebraska Act.

A few Southerners had moved into the northern territories and brought slaves. Other Southerners were concerned when they traveled north with their slaves. Did slaves become free when they were in a free state? No one knew the answer to that question.

One congressman called it an "atrocious crime," another said it was a "gross violation of a sacred pledge." Many Northerners, who hadn't really liked the abolitionists—because they seemed like extremists—now joined their ranks.

But that didn't stop the act from getting passed. In the South there was a feeling of triumph. Alexander Stephens, who guided the bill through the Senate, said, "I feel as if the *Mission* of my life was performed."

He didn't realize that his "mission" would destroy the Whig party and help create a new, Northern party—the Republican party. An Illinois politician, Abraham Lincoln, who had been a Whig, became a Republican. This is what he had to say about the Kansas–Nebraska Act in a speech he gave in Peoria, Illinois:

> *The spirit of [17]76 and the spirit of Nebraska, are utter antagonisms.... Little by little...we have been giving up the old for the new faith....we began declaring that all men are created equal; but now from that beginning we have run down to the other declaration, that for some men to enslave others is a "sacred right of self-government." These principles cannot stand together....Let us re-adopt the Declaration of Independence....If we do this we shall not only have saved the Union: but we shall have so saved it, as to make, and to keep it, forever worthy of saving.*

Lincoln was talking a high road, but a lot of people were sloshing in the mud.

A Positive Good?

Working people—especially the immigrants who were pouring into America—saw slaves as competition. They certainly couldn't work for less pay than the slaves, since slaves weren't paid at all. For them, slavery was a matter only of economics. Others saw it as a moral issue. They said slavery was wrong. But the slave owners didn't agree. They no longer seemed to believe, as they had in Thomas Jefferson's day, that they were trapped in a terrible but necessary system. John Calhoun had convinced them that slavery was a "positive good."

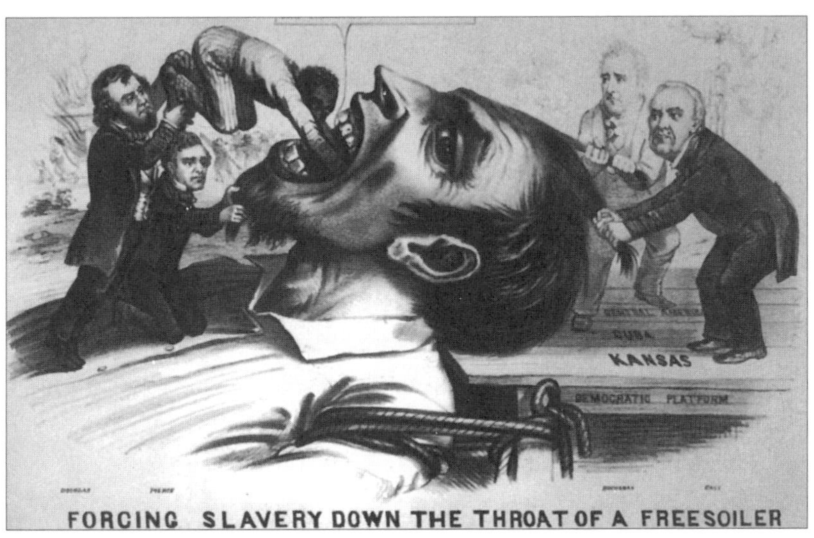

Whichever side won the fight over the territories, one group of Americans would lose out anyway. Who were they?

In this 1858 antislavery cartoon, President James Buchanan and former president Franklin Pierce hold down an antislavery settler while Stephen Douglas tries to shove a helpless slave down his throat. But the free soilers were not about to give in.

A Free State artillery battery in Topeka. They meant business.

Now that popular sovereignty had been declared, Kansans could vote for slavery or against it, according to their beliefs. Missouri Senator David Atchison said, "The game must be played boldly....If we win we carry slavery to the Pacific Ocean." It turned into a nasty game—on both sides.

Slave owners and abolitionists rushed into Kansas. Each group was determined to win the region. Atchison told Jefferson Davis, "We are organizing. We will be compelled to shoot, burn & hang, but the thing will soon be over." Then Atchison took leave from the Senate and led a group of outlaw types (they were called *border ruffians*) into Kansas so they could vote for slavery. They voted. It was illegal, but it didn't seem to matter. Slavery won the first round. A for-slavery delegate was sent from Kansas to Congress.

An ***arsenal*** is a place where firearms are kept.

After that, every time a vote was to be taken, the border ruffians headed for Kansas (with their bowie knives and revolvers). Then the abolitionists began sending their forces; they carried new Sharps breechloading rifles—and they were actually settling in the territory. The free soilers—the antislavery settlers—refused to vote in the rigged elections. They held their own elections. So, in 1856, Kansas had two governments—one for slavery and one against. According to historian McPherson, both sides were "walking arsenals." There were more free soilers living in Kansas than proslavery people—but the ruffians didn't seem to need much encouragement to cross the Missouri border and make trouble.

They were encouraged. One Missouri newspaper wrote, "Let us purge ourselves of all abolition emissaries...and give distinct notice that all who do not leave immediately for the East, will leave for Eternity!"

A posse of some 800 men headed for the free town of Lawrence, Kansas, dragging five cannons with them. They destroyed the place.

While all this was going on in Kansas, back in Washington, D.C., the abolitionist

The town of Lawrence has been sacked, and the fair maid of free Kansas is at the mercy of the "border ruffians"—Pierce and company again. Douglas (right) is scalping an Indian.

senator Charles Sumner stood up in Congress and spoke for two days. Congressional debate usually follows rules of good manners. It makes sense to be polite, even to people you don't like. Sumner didn't consider that. He called the Missourians "murderous robbers" and "hirelings picked from the drunken spew and vomit of an uneasy civilization." That was just for a starter. Then he managed to insult South Carolina's Senator Andrew P. Butler, and he even talked of South Carolina's "shameful imbecility." It was not the kind of speech that could lead to compromise or the working out of problems. But no one expected what happened next.

Stephen Douglas stood and watched as Preston Brooks beat up Charles Sumner (left). Brooks got off with a $300 fine from a Washington judge.

Two days later, Preston Brooks, who was a cousin of Andrew Butler, walked into the Senate. He walked right up to Charles Sumner, who was seated at his desk, and began beating him on the head with a gold-topped cane. Sumner's legs were trapped under the bolted-down desk, and he couldn't move. He was almost killed. He was absent from the Senate for three years after that because of his injuries. The *Richmond* [Virginia] *Enquirer* praised Brooks's action and said, "the vulgar Abolitionists...must be lashed into submission." Brooks received new canes from all over the South.

That wasn't all; there was more to come. This was now a guerrilla war. One of the soldiers in that war was a fierce-eyed abolitionist, a man who had 20 children—his name was John Brown. When he learned what had happened in Lawrence, Kansas, John Brown told his followers: "We must fight fire with fire" and "strike terror in the hearts of the pro-slavery people." When he heard about the caning of Charles Sumner, Brown "went crazy—*crazy*," said someone who knew him. John Brown decided to get even by killing some "slavehounds" near Pottawatomie Creek, where he lived. (No one at Pottawatomie had anything to do with the destruction at Lawrence or the caning of Sumner.)

Brown and seven of his followers kidnapped five proslavery settlers. Then they murdered them with swords. That started things for real. Kansas was now called "Bleeding Kansas." It was proslavery versus no-slavery, and it was a civil war and the first act of a drama that was about to unfold.

John Brown claimed his stare could make a cat slink out of the room. He was a fanatic, but feeling about slavery was by now so intense that many people approved of what he did.

33 A Dreadful Decision

Dred Scott and his wife, Harriet. Scott was freed after his case was over. He got a job as a porter but died in 1858.

What kind of leader was the man who was elected president in 1856? His name was James Buchanan, and one southern senator said, "Even among friends he rarely expressed his opinions." So when he took office, no one was quite sure what to expect. He hadn't said much on the issue of slavery, but Buchanan owed his election to a combination of southern votes and northern bankers' money.

Everyone knew that the Supreme Court was considering a case that dealt with slavery. James Buchanan knew—though it was improper that he did—what the decision would be. So he stood before the nation on his inaugural day, March 4, 1857, and said that slavery was a question that "belongs to the Supreme Court of the United States, before whom it is now pending, and will, it is understood, be speedily and finally settled."

Taney is pronounced TAH-nee.

The slavery issue finally settled—how wonderful that would be! The Supreme Court was led by Chief Justice Roger Taney. Taney had been appointed by Andrew Jackson to replace the renowned John Marshall. In 1857, Taney was 79, and well-respected. So was the court.

An English cartoon of 1856 forecast trouble for the "Disunited States," ripped apart by the ugly figure of slavery.

The Supreme Court has nine justices. Five must agree if a case is to be decided. The justices write an explanation for each decision. That explanation is called an *opinion*. Usually, there is a majority opinion, and, if some justices disagree with the majority, a dissenting opinion. Two days after Buchanan's inauguration, the Supreme Court of the United States issued a decision in the case of *Dred Scott* v. *Sandford*. All nine justices wrote opinions—so different were their views.

Chief Justice Roger Taney

This is what the case was about: Dred Scott

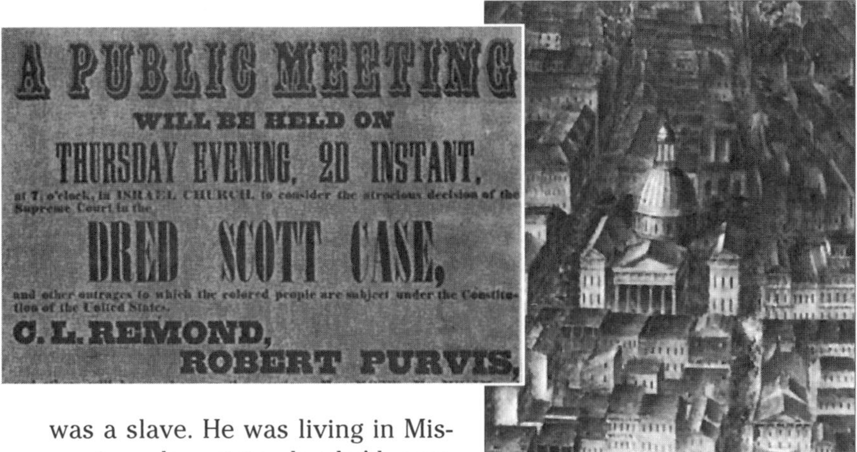

The domed St. Louis courthouse, the first place Scott's case was heard. Taney's decision is considered the worst in the Supreme Court's history.

was a slave. He was living in Missouri—a slave state—but he'd spent several years in Wisconsin—a free territory. Did that make him a free man?

New Yorker John F. A. Sanford (the name of the case was misspelled by the Supreme Court) owned Scott. Sanford was an abolitionist who hated slavery. He could have just freed his slave, but he had bought Dred Scott in order to take a slave case to the Supreme Court. Sanford got a lawyer to help Scott ask the Missouri court for his freedom—saying he should be free because he had lived in a free state. Missouri said Dred Scott couldn't even go to court, because he wasn't a free citizen. The case made its way to the Supreme Court—just as Sanford hoped it would.

Now this is part of what Chief Justice Taney said in his opinion: blacks, even free blacks,

> *had no rights which the white man was bound to respect; and that the Negro might justly and lawfully be reduced to slavery for his benefit.*

Slaves were property, argued Taney, and the Fifth Amendment protects property. Therefore, the Missouri Compromise—which didn't respect the slave owner's property—was *unconstitutional*. (This was going even farther than the Kansas–Nebraska Act.) Wisconsin had not been free territory when Dred Scott lived there, said Taney—even though everyone thought it was.

And, said the court, blacks had no right to citizenship.

That was the decision that President James Buchanan thought would settle the slavery question! What it settled was the question of war. It made it almost certain.

Buchanan

On March 7, 1857, the *New York Daily Times* headline read: "DECISION IN THE DRED SCOTT CASE; the [Northwest] Ordinance of 1787 and the Missouri Compromise Declared Unconstitutional." The same front page gave the menu at President Buchanan's inaugural party. Truffles, venison, mutton, boar's head, and pheasant were served.

A Boxing Champion

Henry Brown had himself packed into a wooden box and mailed to Philadelphia. The box had air holes and was labeled THIS SIDE UP WITH CARE. Despite the label he spent part of the trip upside down—and thought his head would burst. When the crate was finally opened in Philadelphia, Brown stood up and said, "Great God, am I a free man?" He was indeed. For the rest of his life, he was known as "Box" Brown.

34 Fleeing to Freedom

Some slaves had light skin because most of their ancestors were white.

Born to Shame

From Elizabeth Cady Stanton's speech to the American Anti-Slavery Society, published in the abolitionist paper The Liberator*, May 18, 1860.*

To you, white man, the world throws wide her gates; the way is clear to wealth, to fame, to glory, to renown; the high places of independence and honor and trust are yours; all your efforts are praised and encouraged; all your successes are welcomed with loud hurrahs and cheers; but the black man and the woman are born to shame. The badge of degredation is the skin and the sex....For while the man is born to do whatever he can, for the woman and the negro there is no such privilege.

Ellen and William Craft made their freedom journey in December 1848. When they were buying steamship tickets in Macon, Georgia, the steamer's officer look suspiciously at William and said, "Boy, do you belong to that gentleman?"

Ellen Craft was 22, slim, sweet-natured, and shy. She was very intelligent, but she couldn't read or write. She had curiosity, but she had never been far from home. Ellen had white skin, but she was a slave. She would soon be the best-known black woman of her day. Are you confused? Ellen was both white and black. She was biracial (by-RAY-shul). Read on and you will understand.

Ellen was the daughter of James Smith, one of Georgia's wealthiest plantation owners. Smith had a wife and children. Smith had a wife and children. Do you think the typesetter made a mistake and printed the same sentence twice? That was no mistake. Smith had two families. His legal wife was white and mistress of the plantation. The other woman, whom he never married legally, was black and a slave.

Ellen's mother was the slave. Ellen's father treated Ellen and his other slave children like slaves. When Ellen was 11 she was given as a wedding present to her white half-sister. That means she was sent away from the mother who loved her. You can imagine how she felt—she was still a young girl and she was lonely and unhappy. Her half-sister would never admit they were related. Ellen was now a house servant. She learned to be a skilled seamstress; she also learned to be a good listener. She heard the white people talk about the troubles between North and South. She heard that there were people in the North, called abolitionists, who wanted to free the slaves. She decided that she would run away to the North.

Then Ellen fell in love with William Craft, who was a slave and a carpenter. He fell in love with Ellen. William had money saved because his owner allowed him to earn extra money and keep it for himself. They made plans to run away together. They knew that if they were caught they would be beaten, separated, and perhaps sold. It took courage to do what they planned to do.

It was Ellen's idea. She pretended to be a man: a young, sickly white man. She pretended to be journeying to Philadelphia for medical treatment. William was her slave. Because she had listened carefully to her owner's conversations, she knew all about Philadelphia's doctors. But neither Ellen nor William Craft knew how to read a map, or buy tickets for a journey, or sign a name on a hotel register. How could they travel north? How could they carry out their scheme?

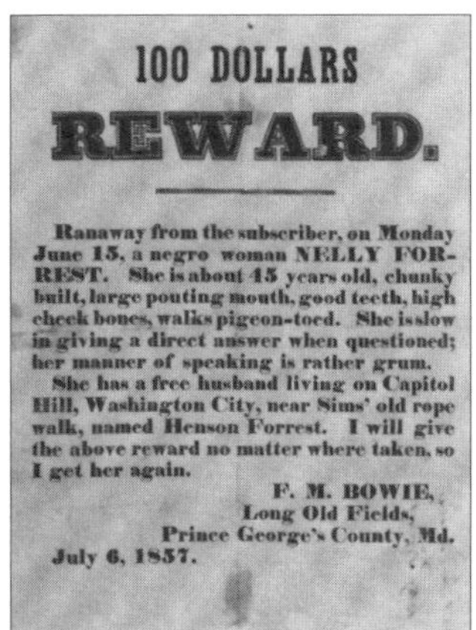

100 DOLLARS

REWARD.

Ranaway from the subscriber, on Monday June 15, a negro woman NELLY FORREST. She is about 45 years old, chunky built, large pouting mouth, good teeth, high cheek bones, walks pigeon-toed. She is slow in giving a direct answer when questioned; her manner of speaking is rather grum.

She has a free husband living on Capitol Hill, Washington City, near Sims' old rope walk, named Henson Forrest. I will give the above reward no matter where taken, so I get her again.

F. M. BOWIE,
Long Old Fields,
Prince George's County, Md.
July 6, 1857.

Less than two years after the Crafts escaped, the Fugitive Slave Law was passed. Now it was a crime to help a runaway slave.

Ellen put her arm in a sling. She said it was injured. That explained why she couldn't write. She put a big bandage around her cheeks to hide her smooth face. No one would wonder why she didn't have whiskers. William bought her a man's suit; he also bought her high-heeled boots to make her seem taller.

Now Ellen had to forget her shyness. She had to act like a slave owner and order William around. She had to buy tickets. She did it. "A ticket for William Johnson and slave," she said in a strong voice. She was William Johnson. Because whites and blacks could not sit together, they rode in separate cars. Ellen was frightened. Then something fearful happened. The man who sat next to her was a white man she knew. Would he recognize her in spite of the man's suit she wore? She pretended to be very sick, groaning answers to his questions. He moved away.

Ellen and William Craft—who were now William Johnson and slave—traveled by train and boat. They stayed in a fine hotel; William Johnson ate in the hotel's dining room. They had more than one

A slave nursemaid might care for the owner's children and see her own children sold away.

Common Destiny

Frederick Douglass, the great abolitionist, had also been a slave, and had run away to freedom. In 1851 he said this to white people:

Have we not a right here? For 300 years or more, we have had a foothold on this continent. We have grown up with you. We levelled your forests. Our hands removed the stumps from your fields and raised the first crops and brought the first produce to your tables. We have fought for this country....I consider it settled that the black and white people of America ought to share common destiny. The white and black must fall or flourish together. We have been with you, are still with you, and mean to be with you to the end. We shall neither die out nor be driven out. But we shall go with you and stand either as a testimony against you or as evidence in your favor throughout all your generations.

Sale of Slaves and Stock.

The Negroes and Stock listed below, are a Prime Lot, and belong to the ESTATE OF THE LATE LUTHER McGOWAN, and will be sold on Monday, Sept. 22nd, 1852, at the Fair Grounds, in Savannah, Georgia. at 1:00 P. M. The Negroes will be taken to the grounds two days previous to the Sale, so that they may be inspected by prospective buyers.

On account of the low prices listed below, they will be sold for cash only, and must be taken into custody within two hours after sale.

No.	Name.	Age.	Remarks.	Price.
1	Lunesta	27	Prime Rice Planter,	$1,275.00
2	Violet	16	Housework and Nursemaid,	900.00
3	Lizzie	30	Rice, Unsound,	300.00
4	Minda	27	Cotton, Prime Woman,	1,200.00
5	Adam	28	Cotton, Prime Young Man,	1,100.00
6	Abel	41	Rice Hand, Eyesight Poor,	675.00
7	Tanney	22	Prime Cotton Hand,	950.00
8	Flementina	39	Good Cook, Stiff Knee,	400.00
9	Lanney	34	Prime Cottom Man,	1,000.00
10	Sally	10	Handy in Kitchen,	675.00
11	Maccabey	35	Prime Man, Fair Carpenter,	980.00
12	Dorcas Judy	25	Seamstress, Handy in House,	800.00
13	Happy	60	Blacksmith,	575.00
14	Mowden	15	Prime Cotton Boy,	700.00
15	Bills	21	Handy with Mules,	900.00
16	Theopolis	39	Rice Hand, Gets Fits,	575.00
17	Coolidge	29	Rice Hand and Blacksmith,	1,275.00
18	Bessie	69	Infirm, Sews,	250.00
19	Infant	1	Strong Likely Boy	400.00
20	Samson	41	Prime Man, Good with Stock,	975.00
21	Callie May	27	Prime Woman, Rice,	1,000.00
22	Honey	14	Prime Girl, Hearing Poor,	850.00
23	Angelina	16	Prime Girl, House or Field,	1,000.00
24	Virgil	21	Prime Field Hand,	1,100.00
25	Tom	40	Rice Hand, Lame Leg,	750.00
26	Noble	11	Handy Boy,	900.00
27	Judge Lesh	55	Prime Blacksmith,	800.00
28	Booster	43	Fair Mason, Unsound,	600.00
29	Big Kate	37	Housekeeper and Nurse,	950.00
30	Melie Ann	19	Housework, Smart Yellow Girl,	1,250.00
31	Deacon	26	Prime Rice Hand,	1,000.00
32	Coming	19	Prime Cotton Hand,	1,000.00
33	Mabel	47	Prime Cotton Hand,	800.00
34	Uncle Tim	60	Fair Hand with Mules,	600.00
35	Abe	27	Prime Cotton Hand,	1,000.00
36	Tennes	29	Prime Rice Hand and Cocahman,	1,250.00

A notice of a slave sale in 1852. "Slaves will be sold separate, or in lots," means that families could be broken up.

close call, but they made it—to Philadelphia and freedom.

When they reached Philadelphia, they found free blacks who sent them to the home of an abolitionist family. But when Ellen saw the people she refused to enter their house. They were white. Ellen didn't know that white people could be nice. She didn't believe that white people would help her. She thought it was all a trick. But she began to find that skin color has nothing to do with kindness or meanness. Those white people became her friends.

Ellen and William Craft were soon famous. People wanted to hear their story. They moved to Boston, which was a center of abolitionism. They made speeches. Articles were written about them. They would have preferred a quiet life, but they understood the need to tell their story to help others who were still enslaved.

When slave catchers came from Georgia just to capture the Crafts, Henry Ingersoll Bowditch, the son of the renowned navigator Nathaniel Bowditch, whom you read about in chapter 14, helped Ellen and William flee to England. In England they learned to read and write. They taught sewing and carpentry. Then some English merchants sent William to Africa to sell their goods. But in Africa the king of Dahomey took his goods and gave him chained slaves in return. Craft would have nothing to do with slavery. He freed the slaves. That was the end of that job.

Ellen and William were homesick. They longed for their own country. Finally the time came (after the Civil War) when it was safe for them to go home. They spent the rest of their lives in the South, teaching and helping others.

When William Johnson and slave walked down that long, winding American road toward freedom and justice, they didn't realize they would be speaking out for all those left behind. They learned that it would take hard work to make the words of the Declaration of Independence mean what they said. Ellen and William Craft were willing to do their part.

Ellen Craft had become prejudiced against white people. Knowing her background, you can understand that. But does bigotry make sense? Ellen was more white than black.

35 Over the River and Underground

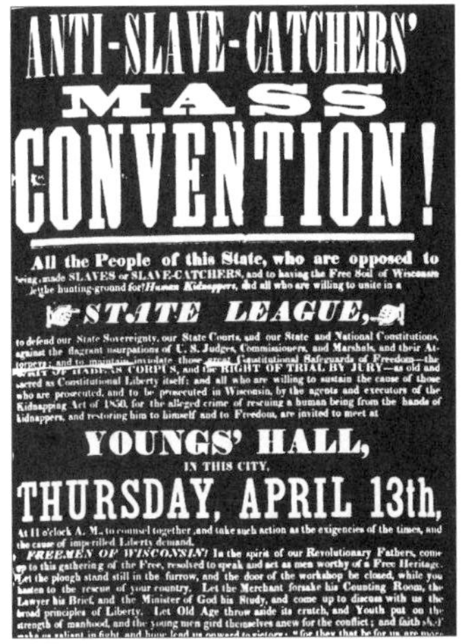

A Wisconsin rallying cry to "All the people of this State, who are opposed to being made slaves or slave-catchers."

John Price was a slave in Kentucky. We don't know much about him—what he thought about or what he was like—but we do know he must have hated being a slave. He was willing to risk his life to run away.

He got a chance to do it one cold winter. It was so cold that the wide Ohio River froze over. Price and two friends —a woman named Dinah and a man named Frank—decided to cross the river. In the dark of night they took two of their master's horses and headed out onto the slick ice. When they got to the Ohio side of the river there was no way to get onto the land. The river's banks were too steep and the horses kept slipping. Soon the three runaways feared they might freeze to death. Then, as morning came, they saw where a road cut through the hilly bank. They had made it to a free state.

But that didn't mean they were free. According to the Fugitive Slave Law, anyone who found them in Ohio was obliged to return them to their owner. If you broke the law you could go to jail, or be fined, or both.

John, Dinah, and Frank were lucky: the first person they met was a Quaker man who would not obey that law. He was willing to risk a jail sentence. He fed them and gave them a place to rest.

The fugitives were heading for Canada, where slavery had been abolished. Before they went on they let their horses go, sending them back toward Kentucky, where their owner found them. Then Dinah started out; she thought it safer to go by herself. No one knows what

Most slaves were Christians and, like the abolitionists, they found a message of fairness and brotherhood in that religion. Southern slaveholders read passages in Christian scripture that told them that slaves should obey their masters. Slaves read the Bible's stories of freedom. Each group found what it wanted in the same Bible.

A station on the Underground Railroad. From 1830 to 1860 about 40,000 fugitives passed through Ohio alone.

happened to her. John and Frank went next. The Quaker man sent them all traveling on the Underground Railroad.

You may have heard of that railroad. Well, it wasn't a real railroad, and it wasn't underground. Still, the name made sense. The Underground Railroad was a secret way of travel, with conductors and stations and passengers. The passengers were blacks escaping from slavery. The conductors—who were black and white—helped them along the way. The stations were places where people could be trusted to feed and house and help the runaways. Some of those places were houses with special hidden rooms; some were barns; some were even riverboats.

The fugitives traveled at night, following the North Star. Sometimes the nights were cloudy and there were no stars to follow. Sometimes hunting dogs were sent to track them down. Usually the passengers traveled through places they had never been before. Often they were hungry. Always they were scared. But the idea of freedom gave them the courage they needed. As they went they whispered the locations of the railroad's stations to one another.

John Price made it to northern Ohio and then he stopped. He didn't think he needed to go farther. He was in Oberlin, which was known as an abolitionist town. It was built around a small college, called Oberlin College, where blacks and whites, men and women, all went to school together. There wasn't another college in the country like that.

Oberlin was different because of the people who founded it. They were mostly Quakers and Presbyterians who thought slavery wrong and against the will of their God. They were people of strong beliefs; most were studying to be ministers. They went out from the college and preached against slavery.

John Price found work as a laborer. He was a pleasant, well-liked man.

Women for Freedom

Anetta M. Lane and Harriet R. Taylor were slaves but that didn't stop them from being women of action. Together with two abolitionists, Joshua R. Giddings and Jollife Union, they formed an active underground railway station in Norfolk, Virginia. Later, when freedom came, they threw off their secrecy and, in 1867, chartered a black women's organization, the United Order of Tents. That group has continued to work for the improvement of conditions among black women nationwide.

One day a boy of 13, Shakespeare Boynton, told Price he knew someone who had a job for him. Shakespeare was not telling the truth. He had been paid to lie to John Price. He took the money and led Price into a trap. He led him to some men who were slave catchers. They had been sent from Kentucky just to catch Price.

John Price was handcuffed and put in a wagon. His captors headed to a nearby town. On the way, their horse and wagon passed two Oberlin men. When Price saw them he shouted out, "Help, help!"

Those men rushed to Oberlin and told others what they had seen. People were outraged. Never had anyone been kidnapped from Oberlin! Blacks and whites hitched their horses to their wagons and galloped off.

In the meantime, Price was in a hotel with his captors waiting for the next train south. Soon the street in front of the hotel was filled with angry people from Oberlin. They recaptured Price and put him on a train for Canada. It was September of 1858, and he was not heard of again.

But that isn't the end of his story. The men who freed him had broken the Fugitive Slave Law. They were in trouble. There was a crowd of them: a doctor, a carpenter, a cabinetmaker, a printer, an undertaker, a

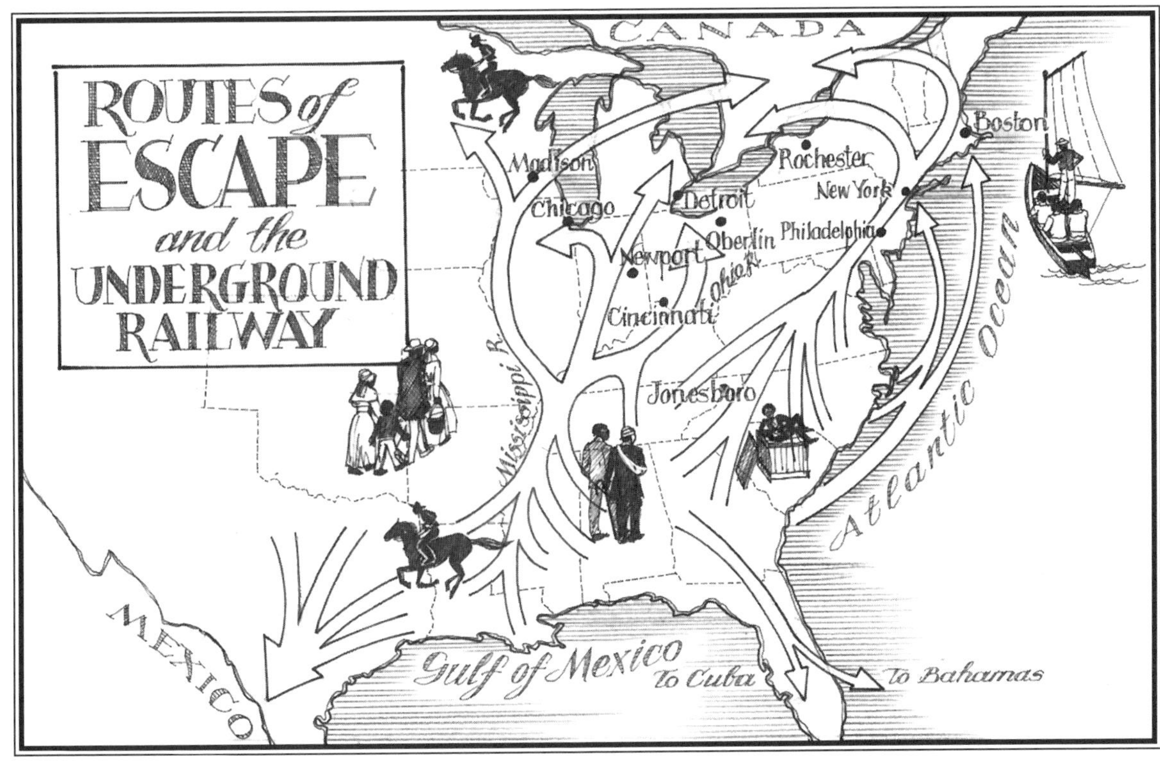

A plantation policeman inspects a slave's off-hours pass. A slave without a pass might be a runaway and would be sent home to get a beating.

schoolteacher, a brickmaker, a lawyer, a grocer, a harnessmaker, some students, three fugitive slaves living in Oberlin, and some farmers.

They had defied the government and its laws. President James Buchanan wasn't going to let them get away with that. The rescue in Oberlin was talked about throughout the country. Those who believed in slavery saw it as a test case. Would the government enforce the Fugitive Slave Law, they asked? The people who were against slavery also saw it as a test case. Would the government send the leading citizens of a community to jail for helping someone to become free? An editorial in the *Ohio State Journal*, an abolitionist paper, said:

> *It is not so much a violation of the Fugitive Slave Law which is to be punished by the United States as the anti-slavery sentiment. That is the thing. It is Oberlin, which must be put down. It is freedom of thought which must be crushed out.*

The trial was big news. I won't go into all the details—it would take too much space—but all the accused men were kept in jail while waiting for their trials. The jailer—who liked them—tried to make them as comfortable as possible. Still, there were rats and roaches in the jail, and, of course, they couldn't go home when they wanted to. But friends were allowed to visit, including all the children in the Sunday-school class that one prisoner taught.

Finally it was time for their cases to come before the court.

Simeon Bushnell, a 29-year-old Oberlin printer who was married and father of a baby, was first to be tried. Bushnell, a quiet-natured working man with no spare money, was short and stocky, with a beard and dark eyes. He was known to be a conductor on the Under-

No Slave Upon Our Land

George Latimer—an alleged runaway slave—was seized in Boston, without a warrant, at the request of James Grey of Norfolk, Virginia, who said he owned Latimer. Fifty thousand Massachusetts citizens signed a petition protesting that action, and John Greenleaf Whittier wrote a long poem, "Massachusetts to Virginia," which was very popular in the North. In the poem, Southerners are called sinners. *How do you think they felt about that? Here are the last three stanzas:*

Look to it well, Virginians! In calmness we have borne
In answer to our faith and trust, your insult and your scorn;
You've spurned our kindest counsels; you've hunted for our lives;
And shaken round our hearths and homes your manacles and gyves!

We wage no war, we lift no arm, we fling no torch within
The fire-damps of the quaking mine beneath your soil of sin;
We leave ye with your bondmen, to wrestle, while ye can,
With the strong upward tendencies and godlike soul of man!

But for us and for our children, the vow which we have given
For freedom and humanity, is registered in heaven;
No slave-hunt in our borders—no pirate on our strand!
No fetters in the Bay State—no slave upon our land!

(What is the Bay State? What are bondmen? Manacles? Gyves? Fetters?)

Setting out from Maryland's Eastern Shore for a Delaware Railroad depot.

ground Railroad. Bushnell was found guilty, given a prison sentence, and fined.

The next person to be tried was a 40-year-old schoolteacher named Charles Langston. Langston, a Virginian, was part black, part white, part Indian. He had devoted much of his life to fighting slavery. Langston,was an officer in the Ohio Anti-Slavery Society and an agent for the Sons of Temperance (people who were against drinking). Charles Langston spoke before the court. This is some of what he said:

> *My father was a revolutionary soldier...he served under Lafayette, and fought through the whole war; and...he always told me that he fought for my freedom as much as for his own....[He taught me] that the fundamental doctrine of this government was that all men have a right to life and liberty.*

Langston asked the judge what he would do if his wife, or child, or brother were taken into slavery. Would the judge resist the laws? Langston believed he would.

> *We have a common humanity. You would do so; your manhood would require it; and no matter what the laws might be, you would honor yourself for doing it; your friends would honor you for doing it; and every good and honest man would say, you have done right!*

The courtroom rang with applause.

It didn't matter. Langston was convicted, sentenced, and fined.

Now it just happened that the citizens of Oberlin were not the only ones who had broken the law. The slave catchers had not taken John Price legally. Slaves couldn't just be kidnapped. The slave catchers had to show the proper papers. That hadn't been done. The Oberlin lawyers got busy. They had the slave catchers arrested.

The slave catchers were terrified of going to jail. They knew the jailers would not be nice to them. They agreed to drop their case if the Oberlin people did the same thing. So, finally, it was all over. Except, of course, it wasn't at all. The issue of slavery hadn't been solved. The conflict was just beginning.

Driving That Train

These words are from the **Reminiscences of Levi Coffin*****, one of the most important organizers of the Underground Railroad.***

One company of 28 that crossed the Ohio River at Lawrenceburg, Indiana—20 miles below Cincinnati—had for conductor a white man whom they employed to assist them....Their plight was a most pitiable one. They were cold, hungry, and exhausted; those who had lost their shoes in the mud suffered from bruised and lacerated feet....They could not enter the city for their appearance would at once proclaim them to be fugitives....I requested friend Cable to keep the fugitives as secluded as possible until a way could be provided for safely forwarding them on their way to Canada. Friend Cable was a stockholder in the Underground Railroad, and we consulted together about the best route....West Elkton, 25 or 30 miles from College Hill, was the first Underground Railroad depot. That line always had plenty of locomotives and cars in readiness. I agreed to send information to that point, and accordingly wrote to one of my particular friends at West Elkton, informing him that I had some valuable stock on hand which I wished to forward to Newport, and requested him to send three two-horse wagons—covered—to College Hill, where the stock was resting.

"Get Off the Track," sang abolitionists in 1844. "The freedom train is coming."

36 Seven Decades

"I am naturally antislavery," said Abraham Lincoln. "If slavery is not wrong, nothing is wrong."

Remember when Charles Thomson galloped to Mount Vernon to tell George Washington he had been elected president? That was in 1789. Remember George Washington's triumphant journey from Virginia to New York? Remember the parades and the cheering?

Boys and girls who were 10 when George Washington was inaugurated are now 81 and gray-haired.

It is 1860, and a man from Illinois, Abraham Lincoln, has won the presidency. There will be no parades for Lincoln and little cheering. Some people are saying that the American experiment in democracy—started so bravely 71 years earlier—is finished. It has lasted longer than anyone expected. But now, they say, the Union is doomed.

How could that have happened? How could the United States, with all its energy and optimism, be in grave danger?

Well, the country had been born with a promise, a paradox, and a problem. It had tried to avoid the problem; good people had hoped it would go away. But problems have a way of getting worse.

The Hiss of Mobs

Governor Sam Houston was impeached by Southern leaders when he refused to withdraw Texas from the Union in support of slavery. He died in 1863. This was what he thought about it:

The voice of the people is not always the voice of God. When selfish political leaders succeed in arousing public prejudice and stilling the voice of reason...the voice of the people then becomes the hiss of mobs....I am an aging man. My hair is turning gray from toiling, as I believe for the liberties of mankind....I regret to say that every day seems to lessen the respect which politicians hold...for the Constitution. ...The Constitution was intended to protect the rights of the humblest citizen. If it is cast aside or trampled down, the citizen has no rights. He must appeal to the mercy of tyrants instead of the laws of the land. These are some of the principles that have always actuated me in life and I hope to cherish them as long as I live.

Sometimes the longer you put them off the harder they are to solve. The problem, of course, was slavery.

And the promise—what was the promise? The promise was in the words of the Declaration of Independence. The great Declaration that had helped form the nation shone like a bright light to people all over the world. Its words said:

> *We believe that all Men are created equal, that they are endowed by their Creator with certain unalienable Rights, that among these are Life, Liberty, and the Pursuit of Happiness—that to secure these Rights, Governments are instituted among Men, deriving their just Powers from the Consent of the Governed.*

A Stand for Peace

"The determination of our slaveholding President to prosecute the war, and the probability of his success in wringing from the people men and money to carry it on, is made evident...by the puny opposition arrayed against him....None seem willing to take their stand for peace at all risks."

Those strong words were written by someone who was willing to take a stand for peace. He was a former slave, and he had the courage to speak out against a popular war (the Mexican War) when most Americans were cheering it. He was Frederick Douglass, and he had his own newspaper, the *North Star*, where he said what he thought and said it well.

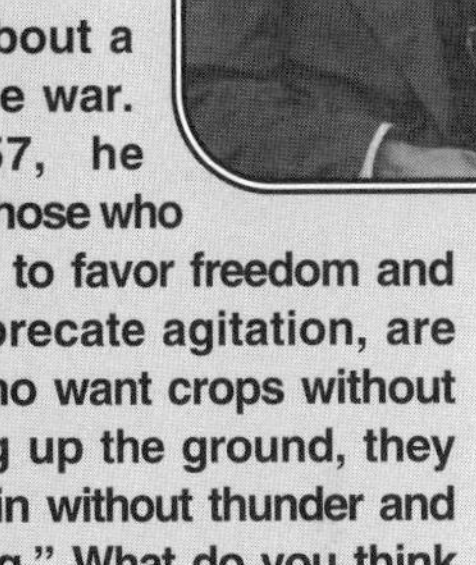

When it came to fighting for freedom for the slaves, Douglass thought differently about a possible war. In 1857, he said, "Those who profess to favor freedom and yet deprecate agitation, are men who want crops without plowing up the ground, they want rain without thunder and lightning." What do you think about his different stands on war and peace?

(Deprecate *means to disapprove;* agitation *means stirring up. In this case Douglass meant to stir up the North against slavery.)*

It was something new for a government to make that kind of promise—equality for all. It was something new and wonderful and special.

And that created the paradox. Because the reality was different from the promise. In America there were people—just like you and me—who lived the lives of prisoners. If they tried to escape, they faced armed patrols and attack dogs. How could men and

This British antislavery cartoon sneers at the hypocrisy of proslavery people's talk of freedom in a land where thousands are not free. The slavemaster, flourishing a lash, hides behind a mask of Liberty.

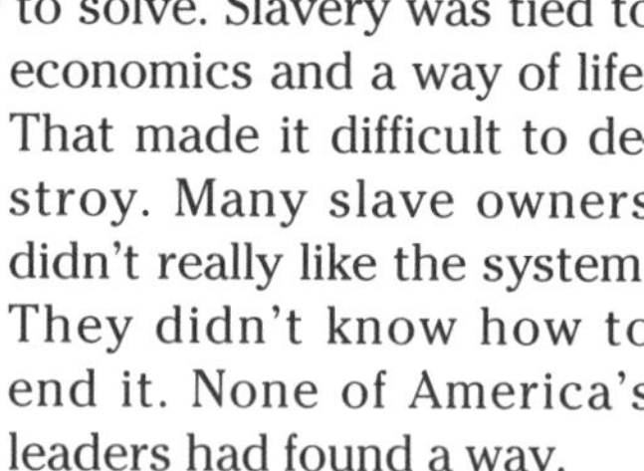

In 1860, Lincoln, "the new Cabinet-maker" has an impossible job to stick North and South back together with "union glue."

women, who cared so much about liberty, keep their brothers and sisters in chains? How could they allow slavery? That was a paradox.

And African-Americans weren't the only ones for whom the promise of the Declaration was not met. Indians, Asians, and women did not have equal rights.

"We didn't mean you," said some of the nation's leaders. "We were only talking about white men being equal," they said.

Fair-minded people began to question that answer.

The problem of injustice is never easy to solve. Slavery was tied to economics and a way of life. That made it difficult to destroy. Many slave owners didn't really like the system. They didn't know how to end it. None of America's leaders had found a way.

Finally it would all come to war. No one wanted it, but there seemed no way to prevent it. It would be the bloodiest of wars—a civil war—with cousin fighting cousin and families torn apart. When it was over, America's childhood was over. The nation would be weary and wounded, but more fair. Slavery would be dead and the Constitution would have three new amendments. Those amendments would make the promise of equality the law of the land.

Big Trouble Ahead

See that skinny man in striped pants standing on the edge of a cliff? His name is "Uncle Sam" and he is a likable guy. But it looks as if he is in big trouble. Some people are chained to him and they are pulling and tugging. If he loses his balance they will all fall over the precipice. (*Precipice* is a fancy word for a cliff.)

Oh, look, as if things aren't bad enough for Uncle Sam, some bad guys are riding hard toward him and their guns are raised. The bad guys—Greed, Cruelty, and Racism—are well-known in these parts. They are powerful outlaws.

Will our hero fall off the cliff and be destroyed? Or will the forces of good—Kindness, Unselfishness, Fairness—come riding to his rescue?

Read on, and you'll see what happens to our uncle. War is coming, a Civil War, and that isn't civil at all.

Chronology of Events

1802: Nathaniel Bowditch publishes *The American Practical Navigator*
1806: Zebulon Pike explores the Southwest
1819: Stephen Long explores the Rocky Mountains
1820: under the Missouri Compromise, Maine becomes the 23rd state and the 12th free state
1821: Missouri becomes the 24th state and the 12th slavery state
1821: William Becknell opens the Santa Fe Trail
1821: Stephen Austin leads 300 settlers to Texas
1828: Andrew Jackson is elected seventh president
1830: the Cherokees and other Indians are forced westward by the Indian Removal Act (see Book 4)
1830: the Baltimore & Ohio Railroad opens the first track for steam locomotives
1831: William Lloyd Garrison publishes *The Liberator*
1832: more than 1,000 men meet for the first rendezvous of William Ashley's mountain men
1832–34: George Catlin paints Indians of the Plains
1833: Jackson defeats the Bank of the United States
1835: the penny press begins with the *New York Sun*
1836: Martin Van Buren is elected eighth president
1836: Arkansas becomes the 25th state
1836: Mexicans defeat Americans at the Alamo
1837: the Grimké sisters lecture on slavery's evils
1837: Michigan becomes the 26th state
1837: Mary Lyon opens Mount Holyoke College for women in Massachusetts
1837: financial panic and then a depression grip U.S.
1838: John James Audubon finishes *Birds of America*
1838: Frederick Douglass flees to Massachusetts
1839: Joseph Cinque captured in Sierra Leone
1840: William Henry Harrison elected ninth president
1841: Harrison dies; John Tyler is 10th president
1841: Manjiro comes to America on a whaling ship
1843: Dorothea Dix begins crusade to reform treatment of mentally ill
1844: first telegraph message sent in Morse code
1844: James K. Polk is elected 11th president
1845: Florida and Texas are the 27th and 28th states
1845: 3,000 pioneers take the Overland trails west
1845: John L. O'Sullivan talks of "manifest destiny"
1846: the Oregon Treaty gives Oregon to the U.S.
1846: Iowa becomes the 29th state
1846–1848: the Mexican–American War
1848: Zachary Taylor is elected 12th president
1848: Wisconsin becomes the 30th state
1848: Brigham Young and Mormons settle in Utah
1848: the Seneca Falls Women's Rights Declaration
1848: John Quincy Adams dies in Congress
1848: Ellen and William Craft escape slavery
1849: the California gold rush begins
1850: Henry Clay, John Calhoun, and Daniel Webster forge the Compromise of 1850
1850: Fugitive Slave Law forbids helping runaways
1850: Taylor dies; Millard Fillmore is 13th president
1850: the National Road reaches Vandalia, Illinois
1850: California becomes the 31st state
1851: women's rights reformers Elizabeth Cady Stanton and Susan B. Anthony meet
1851: Sojourner Truth speaks for women and slaves
1852: more than 20,000 Chinese enter California
1852: death of Henry Clay and Daniel Webster
1852: Franklin Pierce is elected 14th president
1853: Matthew Perry sails into Tokyo Bay, Japan
1854: the Republican party is founded
1854: the Rock Island line crosses the Mississippi
1854: Kansas–Nebraska Act divides western settlers
1854: Henry David Thoreau publishes *Walden*
1855: Walt Whitman publishes *Leaves of Grass*
1856: James Buchanan is elected 15th president
1857: the *Dred Scott* decision
1858: Minnesota becomes the 32nd state
1858: the Oberlin trial tests the Fugitive Slave Law
1858: the Overland stagecoaches reach California
1859: the Comstock Lode is discovered in Nevada
1859: Oregon becomes the 33rd state
1860: the Pony Express begins
1860: Abraham Lincoln is elected 16th president
1861: Rebecca Harding writes on factory conditions

More Books to Read

Here are books that you will like—the kind that keep you up until you finish them. Check with your librarian for other books about this exciting time when Americans were whalers, pioneers, cowboys, slaves, farmers, builders, and more.

Louisa M. Alcott, *Little Women* (many editions available). You will enjoy the story of Meg, Jo, Beth, and Amy as much as girls—and boys—did when this book first appeared in 1868. It is funny and sad. If you like sequels, you can go on to *Good Wives, Little Men*, and *Jo's Boys*—but *Little Women* is the best.

Bess Streeter Aldrich, *A Lantern in Her Hand,* D. Appleton & Co., 1928. Abbie Deal gives up the chance of marrying a fancy doctor and studying singing, and moves from Iowa to Nebraska Territory with her pioneer husband, Will. Together with other settlers, they struggle to plant trees and make farms and build Cedartown out of virgin prairie. This book makes you feel how the Midwest was tamed.

Avi, *The True Confessions of Charlotte Doyle,* Orchard, 1990. It is 1832. Sent home to America from her English school, 13-year-old Charlotte is the only passenger aboard a merchant ship with a cruel captain and a mutinous crew. This exciting book tells about her incredible adventures—and has lots of interesting information about 19th-century sailing ships.

Rachel Baker, *The First Woman Doctor,* Simon & Schuster, 1944. Most biographies of famous people are dull, but much of this story of Elizabeth Blackwell is told in her own words and is very enjoyable.

Carol Ryrie Brink, *Caddie Woodlawn,* Macmillan, 1935. Caddie was a real little girl who grew up in a pioneer community in the Wisconsin woods in the 1860s. Caddie and her brothers and sisters play games and pranks, go to school, do chores. Caddie makes friends with an Indian, hears of President Lincoln's death, and is almost struck by lightning in a storm. I think you will like this book.

Paula Fox, *The Slave Dancer,* Bradbury, 1973. In 1840, Jessie Bollier, aged 13, is kidnapped from his home in New Orleans onto the *Moonlight*, a slave ship bound for West Africa. Jessie's horrible job is to play his fife for the slaves to "dance" to, to keep their muscles from atrophying on the voyage. In the end, Jessie and a young slave are the only survivors of a shipwreck. A marvelous but sometimes horrifying story.

John Jakes, *Susanna of the Alamo,* Harcourt Brace Jovanovich, 1986. This is an easy-to-read picture book that tells the story of Davy Crockett, Santa Anna, and the Alamo from Susanna's point of view.

Alexander Laing, *Seafaring America,* American Heritage, 1974. An adult book, but not hard to read, and full of fascinating details. If you want to find out about the grisly voyage of the whaler *Essex* of Nantucket, look in this book.

Jean Lee Latham, *Carry On, Mr. Bowditch,* Houghton Mifflin, 1955. This splendid book tells the story of Nathaniel Bowditch, the "arithmetic" sailor. It is more like an adventure story than a biography, and it is very good at explaining some of the complicated science behind Bowditch's work.

Julius Lester, *To Be a Slave,* Dial, 1968. A short book that tells a great deal about slavery in real slaves' own words. Some of it is quite hard to bear.

Milton Meltzer, *Underground Man,* Bradbury, 1972. Joshua Bowen was a farmer's boy from Vermont who risked his life as a conductor on the Underground Railroad, guiding escaped slaves out of the South. The author based this novel on the life of Calvin Fairbank, a real abolitionist.

Honoré Morrow, *On to Oregon!,* Morrow, 1946. Halfway along the Oregon Trail in 1843, 13-year-old John Sager and his six brothers and sisters (the youngest a tiny baby) find themselves alone in the world when both their parents die. This is the incredible true story of how John got all of them to the Willamette Valley alive—barely.

Barbara Smucker, *Runaway to Freedom,* Clarke, Irwin, 1977. Julilly and Liza escape their Mississippi plantation and follow the North Star and the Underground Railroad to Canada. This book really helps you feel what it might have been like to run and hide.

Index

Picture Credits

cover: Taylor, *The American Slave Market*, 1852, Chicago Historical Society; **1**: Winslow Homer, *The Cotton Pickers*, 1876, Los Angeles County Museum of Art; **2 (top left)**: Library of Congress; **2-3 (top)**: Thomas Gilcrease Institute of American History and Art, Tulsa; **2-3 (bottom)**: International Center of Photography, George Eastman House, Rochester; **4**: sketch by Daniel Ricketson, 1854; **9 (left)**: Mrs. Frances Trollope, *Domestic Manners of the Americans*; **9 (right)**: American Antiquarian Society, Worcester, Massachusetts; **10**: George Caleb Bingham, *The County Election (1)*, St. Louis Art Museum; **11 (top)**: Gimbels Coin Department; **11 (bottom)**: Library of Congress; **13 (top)**: Independence National Historical Park, Philadelphia; **13 (bottom)**: Library of Congress; **14**: American Philosophical Society, Philadelphia; **15**: Yale University Library; **16 (top)**: Historical Society of Montana; **16 (bottom)**: New York Public Library, Rare Book Division; **17 (top)**: Walters Art Gallery, Baltimore; **17 (bottom)**: Amon Carter Museum, Fort Worth; **18 (top)**: National Park Service, Grand Teton National Park; **18 (inset)**: Montana Historical Society; **19 (top)**: National Portrait Gallery, Smithsonian Institution; **19 (inset)**: Smithsonian Institution Libraries; **19 (bottom)**: American Philosophical Society, Philadelphia; **20 (top)**: State Historical Assocation, Colorado; **20 (bottom)**: Kansas State Historical Society; **21**: Library of Congress; **22**: Yale University Library; **23 (bottom)**: New York Public Library; **24 (bottom)**: Amon Carter Museum; **24-25**: T. Worthington Whittredge, *Santa Fe*, 1866, Yale University Art Gallery, lent by the Peabody Museum; **26**: Missouri Historical Society; **27**: *Harper's Weekly*, August 13, 1859; **28 (left)**: Library of Congress; **28-29 (top)**: Walters Art Gallery, Baltimore; **30**: Museum of Fine Arts, Boston, M. and M. Karolik Collection; **31 (top)**: Smithsonian Institution; **32**: Walters Art Gallery, Baltimore; **33**: National Museum of American Art, Smithsonian Institution; **34-35**: Denver Public Library, Western Collection; **36**: Henry Huntington Library, San Marino; **37**: Denver Public Library, Western Collection; **38**: Peter Rindisbacher, *Colonists on the Red River in North America*, c. 1825, Public Archives of Canada, Ottawa; **39**: City Art Museum, St. Louis; **40 (top)**: National Museum of American Art, Smithsonian Institution, bequest of Helen Huntington Hall; **40 (bottom)**: Library of Congress; **42 (top)**: Amon Carter Museum, Fort Worth; **42 (bottom)**: Henry Huntington Library, San Marino; **43**: National Portrait Gallery, Smithsonian Institution; **44 (left, right top)**: Yale University Library; **44 (bottom left)**: Library of Congress; **45 (bottom)**: Christensen Family Memorial gift, Brigham Young University, Provo, Utah; **46 (bottom)**: Smithsonian Institution Libraries; **48 (top)**: New York Public Library Picture Collection; **48 (bottom)**: Oakland Art Museum; **49 (right)**: Harry T. Peters, Jr., Orange, Virginia; **49 (left)**: New York Public Library Picture Collection; **51**: New-York Historical Society; **52**: Museum of the City of Solothurn, Switzerland; **54**: Texas Memorial Museum, San Antonio; **55 (top)**: James Walker, *Vaqueros Roping Horses in a Corral*, 1877, Thomas Gilcrease Institute of American History and Art, Tulsa; **55**: San Jacinto Museum of History Association; **56 (top)**: White Memorial Museum, San Antonio; **56 (bottom), 57 (left)**: Alamo Museum, San Antonio; **57 (bottom)**: Daughters of the Republic of Texas Library at the Alamo; **58 (top)**: Archives Division, Texas State Library; **59 (right)**: Daughters of the Republic of Texas Library at the Alamo; **58 (bottom)**: American Antiquarian Society, Worcester; **59**: collection of Mrs. Bill Arthur and Mrs. Al Warner; **60**: National Archives; **61 (top)**: National Academy of Design, New York; **61 (bottom)**: San Jacinto Museum of History Association; **62–63 (bottom)**: Office of Military History, Department of the Army; **63 (top)**: Library of Congress; **64**: Franklin Delano Roosevelt Library; **65 (top)**: California State Library; **65 (bottom)**: Peter E. Palmquist; **66 (bottom)**: Henry Huntington Library, San Marino; **68 (top)**: Zelda Mackay Collection, Bancroft Library, University of California; **68 (bottom)**: Amon Carter Museum; **69 (top)**: Montana Historical Society; **69 (bottom)**: Chinese Women of America Project, Chinese Culture Foundation; **70**: Harry H. Baskerville, courtesy Robert B. Honeyman, Jr.; **71 (top)**: New-York Historical Society; **71 (bottom)**: University Archives, Bancroft Library, University of California; **72 (top)**: California State Library; **72 (bottom)**: Mark Twain Papers, Bancroft Library, University of California; **73 (top)**: National Archives; **73 (bottom)**: Library of Congress; **74 (top)**: Pony Express Museum; **74 (bottom)**: Levi Strauss and Company; **75 (top)**: Thomas Gilcrease Institute of American History and Art, Tulsa; **76 (bottom)**: Esmark Collection of Currier & Ives, New York City; **77 (bottom)**: New York Public Library; **78 (top left)**: National Museum of American History, Smithsonian Institution; **78 (top)**: collection of Mrs. Joseph Carson, Philadelphia; **78 (bottom)**: Thomas Gilcrease Institute of American History and Art, Tulsa; **79 (top)**: Deadwood-Cheyenne State Concord coach, c. 1850, Buffalo Bill Historical Center, Cody, Wyoming; **79 (bottom)**: Library of Congress; **80**: International Museum of Photography, George Eastman House, Rochester; **81 (top)**: Joslyn Art Museum, Omaha; **81 (top and bottom left)**: National Portrait Gallery, Smithsonian Institution; **81 (right)**: transfer from the National Collection of Fine Arts to the National Portrait Gallery, Smithsonian Insitution, gift of the Friends of the National Institute, 1859; **82**: Chicago Historical Society; **82–83**: Walters Art Gallery, Baltimore; **83 (inset)**: Amon Carter Museum; **83 (right)**: National Portrait Gallery, Smithsonian Institution; **84**: Essex Institute; **85 (top)**: Peabody Museum of Salem, gift of East India Marine Society; **85 (bottom)**: Nathaniel Bowditch, *New American Practical Navigator*, 1802; **86-87**: Peabody Museum of Salem; **88 (inset)**: Massachusetts Historical Society; **88 (bottom)**: Library of Congress; **89**: Bourne Whaling Museum, New Bedford; **90 (top left and right)**: Mystic Seaport Museum; **90 (bottom)**: Mariner's Museum; **91 (top left and right)**: Mystic Seaport Museum; **91 (bottom)**: Bourne Whaling Museum, New Bedford; **93**: *Harper's Weekly*, June 1875; **94**: Barber, *Massachusetts Historical Collection*, 1839; **95 (top)**: San Francisco Maritime Museum; **95 (bottom)**: Hart Nautical Museum, Massachusetts Institute of Technology; **96–98 (top, bottom right)**: Hiroshi Nakahama; **98 (bottom)**: *Voyager to Destiny*, courtesy Bobbs-Merrill Company; **103 (top)**: I. Pranishnikoff, *Harper's Weekly*, April 20, 1878; **103 (bottom)**: National Portrait Gallery, Smithsonian Institution; **105 (left)**: Historical Society of Pennsylvania; **105 (top)**: *Gleason's Pictorial Drawingroom Companion*, December 25, 1852; **105 (oval)**: New York Public Library Picture Collection; **106**: Kansas State Historical Society; **107 (top)**: Foster, *Practical Penmanship*, 1832, Dartmouth College; **107 (bottom), 108 (top)**: Kansas State Historical Society; **108 (bottom)** Nebraska State Historical Society; **109 (top)**: Winslow Homer, *Fall Games—the Apple Bee*, *Harper's Weekly*, November 26, 1859, Winslow Homer Collection, Bowdoin College; **110 (top)**: Chicago Historial Society; **111**: St. Louis Mercantile Library Association; **112 (top)**: collection of Leonard B. Huber; **113 (top)**: J. Cornelius Rathbone; **113 (bottom)**: *Every Saturday*, 1871; **114 (top)**: St. Louis Art Museum; **115 (top)**: American Antiquarian Society, Worcester; **115 (bottom)**: New England Council, Connecticut Development Commission; **116 (top, inset)**: Chicago Historical Society; **116 (left and right bottom)**: Mount Holyoke College; **117 (top)**: New-York Historical Society; **118 (bottom)**: Library of Congress; **119 (top left)**: Jenkins, *Art of Writing*, 1813; **119 (right top)**: New York Public Library Picture Collection; **119 (right bottom)**: New-York Historical Society; **120 (left, right)**: St. Louis Art Museum; **120 (bottom)**: New York Public Library Picture Collection; **121 (top)**; Metropolitan Museum of Art, gift of I. N. Phelps Stokes and Hawes family; **121 (bottom)**: New-York Historical Society; **122 (bottom)**: National Library of Medicine; **123 (top)**: National Portrait Gallery, Smithsonian Institution; **123 (bottom)**: Museum of the City of New York, collection of Emily Crane Chadbourne; **124**: National Portrait Gallery, Smithsonian Institution; **125**: National Library of Medicine; **126 (top left)**: National Collection of Fine Arts, Smithsonian Institution; **126 (bottom left)**: National Portrait Gallery, Smithsonian Institution; **126 (top right)**: New-York Historical Society; **127 (top)**: *Harper's Weekly*, 1859; **127 (inset)**: Library of Congress; **128**: Harry T. Peters Collection, Museum of the City of New York; **129**: Denver Public Library, Western Collection; **130**: National Portrait Gallery, Smithsonian Institution; **131**: Metropolitan Museum of Art, gift of I. N. Phelps Stokes and the Hawes family; **133 (top)**: *Leslie's*; **134 (top)**: International Center of Photography, George Eastman House, Rochester; **134 (bottom)**: American Iron and Steel Institute, Washington, D.C., historical file; **135**: International Center of Photography, George Eastman House, Rochester; **136**: H. Lawrence Hoffman; **137 (top)**: International Center of Photography, George Eastman House, Rochester; **137 (bottom)**: Library of Congress; **138 (top)**: Baker Library, Harvard Business School; **138 (bottom)**: Brown Brothers, Sterling, Pennsylvania; **140**: artist unknown, *Little Laborers of New York City*, c. 1890; **141**: International Center of Photography, George Eastman House, Rochester; **141**: Svenska Porträttärkivet, National Museum, Stockholm; **142 (top)**: Metropolitan Museum of Art; **142 (middle left and right)**: New York Public Library Picture Collection; **142 (bottom)**: Metropolitan Museum of Art, gift of I. N. Phelps Stokes and the Hawes family; **143 (top)**: Society for the Preservation of New England Antiquities, Boston; **143 (right)**: J. N. Mead, Harvard class of 1851; **144**: New York Public Library Picture Collection; **145 (bottom)**: Daniel Ricketson, 1854; **146**: Berkshire Athenaeum; **147 (top)**: Shelburne Museum; **147 (bottom left)**: National Portrait Gallery, Smithsonian Institution; **147 (bottom right)**: New York Public Library Picture Collection; **148 (top)**: National Collection of Fine Arts and National Portrait Gallery, Smithsonian Institution; **148 (bottom)**: J. Paul Getty Museum; **149 (top)**: Walt Whitman, *Leaves of Grass* (first edition), New York Public Library; **149 (bottom left)**: unidentified artist, *Mike Fink the Ohio Boatman*, *Davy Crockett's Almanack*, 1838; **149 (bottom right)**: William Gropper, 1939, A.C.A Galleries on loan to Metropolitan Museum of Art; **150-152**: New-York Historical Society; **153 (top)**: National Museum of American Art, Smithsonian Insitution; **153 (bottom left)**: National Gallery of Art, Paul Mellon Collection; **153 (bottom right)**: Thomas Gilcrease Institute of American History and Art, Tulsa; **154 (left)**: Rembrandt Peale, *Rubens Peale with a Geranium*, 1801, Mrs. Norman B. Woolworth; **154 (right)**: Charles Willson Peale, *The Artist in His Museum*, Pennsylvania Academy of Fine Arts, Joseph and Sarah Harrison Collection; **155 (left)**: Joshua Johnston, *Young Lady on a Red Sofa*, Los Angeles County Museum of Art; **155 (right)**: John Singleton Copley, *Watson and the Shark*, 1778, Museum of Fine Arts, Boston, gift of Mrs. George von Lengerke Meyer; **156 (top left)**: collection of the Brick Store Museum, Kennebunk, Maine; **156 (top right)**: George Caleb Bingham, *Fur Traders Descending the Missouri*, 1845, Metropolitan Museum of Art, Morris K. Jessup Fund, 1933; **156 (right inset)**: St. Louis Art Museum; **156 (right bottom)**: New York State Historical Association, Cooperstown; **156 (right bottom inset)**: Museums at Stony Brook, Long Island; **157**: New Haven Colony Historical Society; **158 (bottom)**: Eileen Tweedy, the Albany; **160 (top)**: Chicago Historical Society; **160-161 (bottom)**: Barber, *A History of the Amistad Captives*, 1840; **161 (top)**: New York Public Library Picture Collection; **162 (top)**: National Portrait Gallery, Smithsonian Institution; **163 (left)**: Library of Congress; **163 (top right)**: *Pictorial History of the Confederacy*, 1951; **163 (bottom)**: Underwood and Underwood; **164 (top)**: New York Public Library; **165 (top)**: Metropolitan Museum of Art; **165 (bottom)**: New-York Historical Society; **166 (top)**: National Portrait Gallery, Smithsonian Institution; **169**: J. L. Magee, Library of Congress; **170 (top)**: Kansas State Historical Society; **170 (bottom)**: Library of Congress; **171 (top)**: New York Public Library; **171 (bottom)**: Chicago Historical Society; **172 (top)**: National Park Service, U.S. Department of the Interior; **172 (bottom right)**: Library of Congress; **173 (inset)**: Langston Hughes Memorial Library, Lincoln University, Pennsylvania; **173 (top)**: Missouri Historical Society; **173 (right)** Berkshire Museum; **173 (bottom)**: Henry Huntington Library, San Marino; **174**: Schomburg Center for Research in Black Culture, New York Public Library; **175 (top)**: Chicago Historical Society; **175 (bottom)**: Library of Congress; **176**: Chicago Historical Society; **177**: State Historical Society of Wisconsin; **178**: Cincinnati Art Museum, Cincinnati, Ohio; **180**: Museum of the Confederacy, Richmond, Virginia; **181**: American Antiquarian Society, Worcester; **182**: Chicago Historical Society; **183 (top)**: Library of Congress; **183 (bottom)**: Sophia Smith Collection, Smith College; **184 (top)**: American Antiquarian Society, Worcester; **184 (bottom)**: New York Public Library Picture Collection.

A Note From the Author

You may remember Tamara Glenny. (I told you about her in Book 4 of *A History of US.*) Tamara is the editor of these books and she is from England. I think she is irked with me, and I can't imagine why. I did tease her a bit about King George III, and when I added the War of 1812 she growled something about American bragging. Then she brought up England's King Arthur and Robin Hood, too.

Robin Hood? That fellow who hung around Sherwood Forest? I didn't think it was bragging when I told Tamara that her Hood guy just lived in the forest, but when our Paul Bunyan woke up and stretched his arms, he leveled forests. "Why, if Paul Bunyan ever went to the British Isles," said I to Tamara, "and he put one foot in England and the other in Ireland, the whole place would sink into the ocean."

Do you think that was bragging? I say it's just telling the truth, which—I reminded Tamara—is what historians are supposed to do. Then she quoted Arnold Toynbee, a famous English historian who said, "I find it hard to have patience with historians who boast, as some modern Western historians do, that they keep entirely to the facts of history and don't go in for theories."

Now Toynbee may go in for theories, but I try to stick to the facts. And I was astonished to find that Tamara didn't know all the facts about Paul Bunyan. She didn't know that Slim, who was cook in Paul's logging camp, made flapjacks on a grill so big that to keep it greased they hired seven boys, strapped hams to their feet, and kept them skating.

Well, I could see that I had to educate Tamara about Paul. I did. She now knows that it was Paul who actually dug the St. Lawrence River. (That was while he was trying to tame a whale.) Talking about rivers made me remember Mike Fink and that I'd forgotten to include an important story about him in this book.

Mike was a Pittsburgh boy who ended up as one of William Ashley's mountain men, but, in between, he was a riverboatman who had some amazing adventures.

It was the day he decided to jump the Mississippi that should be in all the history books. He'd been planning it for a while, and when the day came, everyone in Cairo, Illinois, which is on the eastern side of the river, was lined up to watch. Mike had his usual breakfast (a dozen eggs and a side of bacon). Then he geared up and he churned up and he charged down Cairo's main street. And then—horrors—a little girl, a toddler, wandered right out into the street and in his way. What should he do? Mike hesitated a nanosecond. If he ran into her—well, she'd be wiped away—but Mike was a gentleman, so he swerved a bit and then he was on the river's edge and leaping into the air and on his way over the Mississippi. But when he got about three-quarters of the way over—and could actually see the other shore—he realized he wasn't going to make it. That little gentlemanly hesitation, to miss the child, had gotten him off stride. Well, he looked down at Old Muddy and she was roiled up—he really didn't want to have to swim. So, right then, quick as an eye blink, he turned around and headed back to Cairo. Almost didn't make it. He had to angle out, tuck his head in, and cut all the wind's resistance. Even then, he landed with one foot in the river.

Now some folks say, "You see, he was no big deal." Tamara's been reminding me that Mike Fink couldn't even jump the Mississippi. But I say three-quarters over and three-quarters back was pretty good. What rivers did Robin Hood jump?

A NOTE FROM TAMARA GLENNY:
I am writing this from Paul Bunyan's logging camp in Bemidji, Minnesota. We have had to move our editorial offices out here because Joy Hakim does so much rewriting that it is the only place that can supply us with enough paper. And if Joy doesn't stop telling you that I can't take a joke, I may move back to England, where they don't have tall tales—as our Alice in Wonderland will tell you.